PROCEEDINGS OF THE THIRD **BWEA** WIND ENERGY CONFERENCE

HELD AT CRANFIELD APRIL 1981

Organised by
BRITISH WIND ENERGY ASSOCIATION

Published by
BHRA Fluid Engineering

Editor: Dr. P.J. Musgrove

The Organisers are not responsible for statements or opinions made in the papers.

The papers have been reproduced by offset printing from the authors' original typescripts and illustrations to minimise delay. Whilst every effort is made to produce a reasonable volume of preprints the organisers are unable to accept responsibility for the quality of the printing.

When citing papers from this volume the following references should be used:

Title Author Paper No. Pages, 3rd British Wind Energy Association Wind Energy Conference Cranfield, U.K. April, 1981

British Library Cataloguing in Publication Data

BWEA Wind Energy Conference *(3rd: Cranfield: 1981)*

Proceedings of the Third BWEA Wind Energy Conference.
1. Wind power - Congresses
I. Title II. Musgrove, P.J.
621.45 TK1541
ISBN 0-906085-56-X

Published by
BHRA FLUID Engineering
Cranfield, Bedford MK43 0AJ, UK

ISBN 0 906085 56 X

Front Cover: Artist's Impression of 20m Wind Turbine for Orkney.

CONTENTS

The Third British Wind Energy Association Wind Energy Conference was organized by the BWEA.

Conference organising committee:

Dr. P. Musgrove
Dr. J. C. Dixon
Dr. L. Freris
Dr. J. Garside
Dr. R. H. Taylor

Committee of the British Wind Energy Association, 1980-1981

Chairman:	Dr. P. Musgrove University of Reading
Vice-Chairman:	Dr. D. T. Swift-Hook CERL
Secretary:	Dr. E. Mowforth University of Surrey
Treasurer:	Dr. D. Lindley Taywood Engineering
Membership Secretary:	Dr. J. C. Dixon Open University
'Windirections' Editor:	Mr. W. Grylls Windtech Ltd.
Press Officer:	Dr. R. H. Taylor CEGB Planning Dept.
Committee Members:	Mr. M. Anderson Cavendish Laboratory
	Mr. P. Fraenkel ITDG
	Dr. L. Freris Imperial College
	Mr. B. Holmes P.I. Specialist Engineers
	Prof. N. Lipman Rutherford Laboratory/Reading University
	Mr. G. Pontin Wesco Windmills
	Mr. D. F. Warne ERA Technology

Readers wishing to contact members of the Organising Committee or officers of The British Wind Energy Association will find their addresses at the back of this volume.

Lecture by Dr. F.J.P. Clarke to Third Wind Energy Workshop

Cranfield, 9th April 1981

Our National Wind Energy Programme

1. At your two previous conferences, Len Bedford and I have reminded you of the aims of the Department of Energy's programme on wind energy. But because these aims and their implications seem not to be fully understood I thought it worthwhile starting again this year by reminding you of them.

2. The Department's basic aim in the national wind energy programme is to develop the technology in order to define its economic potential and benefits in large-scale electricity generation; and in so doing to help commercialisation. This type of aim is in contrast to that of some other countries, especially the US, in which a target is set for RD&D expenditure aimed at having so many GW's of wind power on stream by a certain date.

3. This difference in approach has various implications. First of all, the UK programme seems rather modest against that initiated in the US under the Carter Administration (though this programme is now being very substantially cut back under the new Administration). This in turn gives rise to the complaint sometimes heard from UK industry that it has rather less Government support for industrial developments than its competitors in other countries.

4. The basic reason for this apparently more tentative approach is that the UK energy policy is based upon coal, conservation and nuclear technologies - the so-called 'coconuc' strategy. These are all proven technologies. In contrast, the potential and reliability of wind power technology still remains to be proved and therefore Government has no firm policy for introducing it, only a policy of research aimed at determining whether it should be introduced.

5. There is no doubt in my mind that as a nation we are also influenced in the way we view these renewable technologies by the relative abundance of our energy resources compared with some of our neighbours. Substantial reserves of oil, gas and coal, when added to the historical development of our nuclear capacity, make us feel more relaxed vis-a-vis the renewable technologies than some of our neighbours who have little or no indigenous resources. For this reason there is no incentive for the UK to mount any sort of 'crash' programme. Nevertheless, there is a public and political wish to have a serious programme to investigate the potential of these resources. What I want to do this morning is to argue that we in the UK do have a serious programme and, moreover, one that over the past year has gained considerably in size and coherence.

6. In reviewing our national programme the Wind Energy Steering Committee has divided it into the different areas listed in Table 1.

7. Areas 1 and 2, the 250 kw and 3 MW horizontal axis machines are now well publicised. Orkney has been chosen as the site for the first machine because the northern islands represent the first real UK market for wind machines. It is good marketing practise to tackle first the high-price low-volume part of the market that gives industry its point of entry for wider developments. The Departments of Energy and Industry have been collaborating closely in these developments with an industrial consortium of companies. The 250 kw

machine is seen as having potential industrial application to overseas markets and hence the DoI has offered a grant towards its construction and commissioning. On the other hand this machine is, in many respects, an intermediate scale version of the 3MW design and therefore the Department of Energy has contributed towards the costs of monitoring its performance and analysing the results. The Department of Energy has taken the lead on the 3MW machine with some financial backing from the NSHEB. The project is to design, contract, commission and monitor the performance of this large machine.

8. The Steering Committee has recognised right from its first meeting that vertical axis designs provide a promising alternative design concept that merits investigàtion. Some work has already been done in North America on the Darrieus design of vertical axis machine; our basic philosophy in the UK is to understand sufficient of this overseas work to evaluate its potential. In order to give us some feel for this type of machine, two 5-10 kw versions are being built at the University of Newcastle; one is already running and another is being designed. But it is not our intention to duplicate at large scale work that has been carried out elsewhere.

9. A second consortium of companies has decided to develop an alternative UK design of vertical axis machine - the Musgrove design. To begin with, a 100 kw version is being designed with financial assistance from the Department of Energy. The further intention is that this should be developed into an intermediate size of machine which, if successful, could lead the large-scale trials, ultimately off-shore and perhaps in the context of international collaboration. Work on the 100 kw machine is now well advanced and the industrial consortium have already begun discussing with us the further plans on the intermediate size of machine.

10. Whatever the outcome of these two UK developments, spearheaded by British companies, we cannot afford to ignore the important work going on overseas, especially in Europe and the USA. Therefore our programme area number 5 is concerned with advance designs of horizontal axis machines that will take account of such overseas developments. The suggestion has been made that we ought to have some form of national test bed where advance design features could be tried out; the Steering Committee has this in mind as a development that should be considered for later years.

11. Programme area 6 is concerned with off-shore siting and with determining, by about mid-decade, which type of machine (horizontal axis or vertical axis) is best suited for this purpose. The advantages and disadvantages, technical and economic, of off-shore machines is also the subject of programme area number 7. The UK is among the countries that are taking the lead in this area and again we would hope to benefit by collaborating in international developments.

12. Programme area 8 is about all aspects of the resource base - wind data, topography, wake-wake interactions etc. We start off in all these renewable technologies by thinking that there should be an adequate information base available on the resource e.g. meteorological data. In every case we find out that there is not and that a substantial programme is needed. We also need to know what density of wind machines is possible (hence the wake studies) and which geographical areas are most favoured on account of environmental and other factors. All these features are included in our programme.

13. Programme areas 9 through to 13 deal with generic studies i.e. studies that have importance to several or perhaps all designs of machine. The only point I should like to make on these areas is that the Chief Scientist has said that he would increasingly like to see the results of such work being taken up by the device teams, perhaps by universities and others working closely with the teams under sub-contracts. Some work will remain generic, but his intention is to encourage the work to move out of the universities and research laboratories into industrial application.

Financial Aspects of the UK's Wind Energy Programme

14. This then is the technical base of our programme. I said earlier that the programme was now expanding quite rapidly and that is shown by looking at Table 2 showing in rough terms the historical expenditure, the expenditure that we are planning in the current financial year, and that we guess we might be asking for in the year beyond that. To date we have never quite achieved the expenditure forecast in any year though as the programme gets under way and as our procedures get streamlined our forecasts are getting more accurate. These planned and guessed figures should not therefore be seen as written in tablets of stone but rather the best broad view of those managing the programmes.

General

15. I should like to make some general points. I have in this talk repeatedly emphasised the international dimension. It would be silly if we as a nation did not take account in our own programmes of things going on overseas. Hence in many areas - in the Darrieus design of machine, in advanced horizontal axis machines, in large vertical axis machines, in off-shore studies and in resource studies such as wake-wake interactions - we see it potentially to our advantage to try to understand what others are doing with a view to working with them in appropriate cases. This international collaboration should not be looked upon as something that is only for Government to initiate. Firms have an important role to play through normal commercial access and licensing agreements and the Department has always encouraged firms to operate in this way. Now that our UK programme is taking off, UK firms should also have something to offer overseas firms in such commercial arrangements. This need to learn from overseas programmes is especially true in matters such as stiff versus compliant machines. In a recent visit that the Chief Scientist and I paid to the US it was put to us quite forcibly that eventually we in the UK would probably have to turn to compliant designs in our own machines if we were to optimise them in terms of cost-effectiveness. We need to think carefully about such issues. We also saw something of the work going on on wood as a blade material. Again it is to be hoped that our own industrial teams will take full account of these overseas developments in proposals that are put forward to Government. On another point, in our UK programme we have assumed that in terms of national energy supplies 'big is beautiful'. However, I have seen no rigorous study proving that this is true. While I believe that the optimum size of machine lies somewhere around a few MW's, this is clearly an assumption which needs a more rigorous investigation. We would look very foolish if at the end of the day it turned out that the advantages of scale in terms of volume production outweighed the advantages of scale in terms of size.

16. Finally I must refer to one very important development in the past year that I have not yet mentioned viz. the stated intention of the CEGB to purchase a large machine as soon as a proved design is commercially available. This is the second important market signal of the past year. Some of our companies feel it is a pity that the design that will win the order will probably have to be of US origin. On the other hand, the CEGB is a commercial undertaking

that has to purchase the best products from wherever they are available. Glyn England will no doubt be referring to this matter later this evening and that will be an excellent opportunity to learn at first hand of the Board's strategy in this area.

17. Our UK programme on wind energy may not be as large as that in some other countries. Nevertheless I hope the facts speak for themselves that it is a growing and serious programme that meets the UK's needs and that is well able to take its place in making an important contribution on the international scene.

* * *

TABLE 1

DEPARTMENT OF ENERGY PROGRAMME

1. 250 kW, HA*, ORKNEY
2. 3 MW, HA, ORKNEY
3. 5-10 kW DARRIEUS
4. MULTI MW VA*
5. ADVANCED DESIGN OF VA
6. OFFSHORE SITING HA AND VA
7. OFFSHORE MACHINES
8. RESOURCE BASE
9. GENERIC STUDIES - MATERIALS
10. GENERIC STUDIES - AERODYNAMICS
11. GENERIC STUDIES - GENERATION AND TRANSMISSION
12. GENERIC STUDIES - STRUCTURES
13. GENERIC STUDIES - ENVIRONMENT

* HA - Horizontal Axis

VA - Vertical Axis

TABLE 2

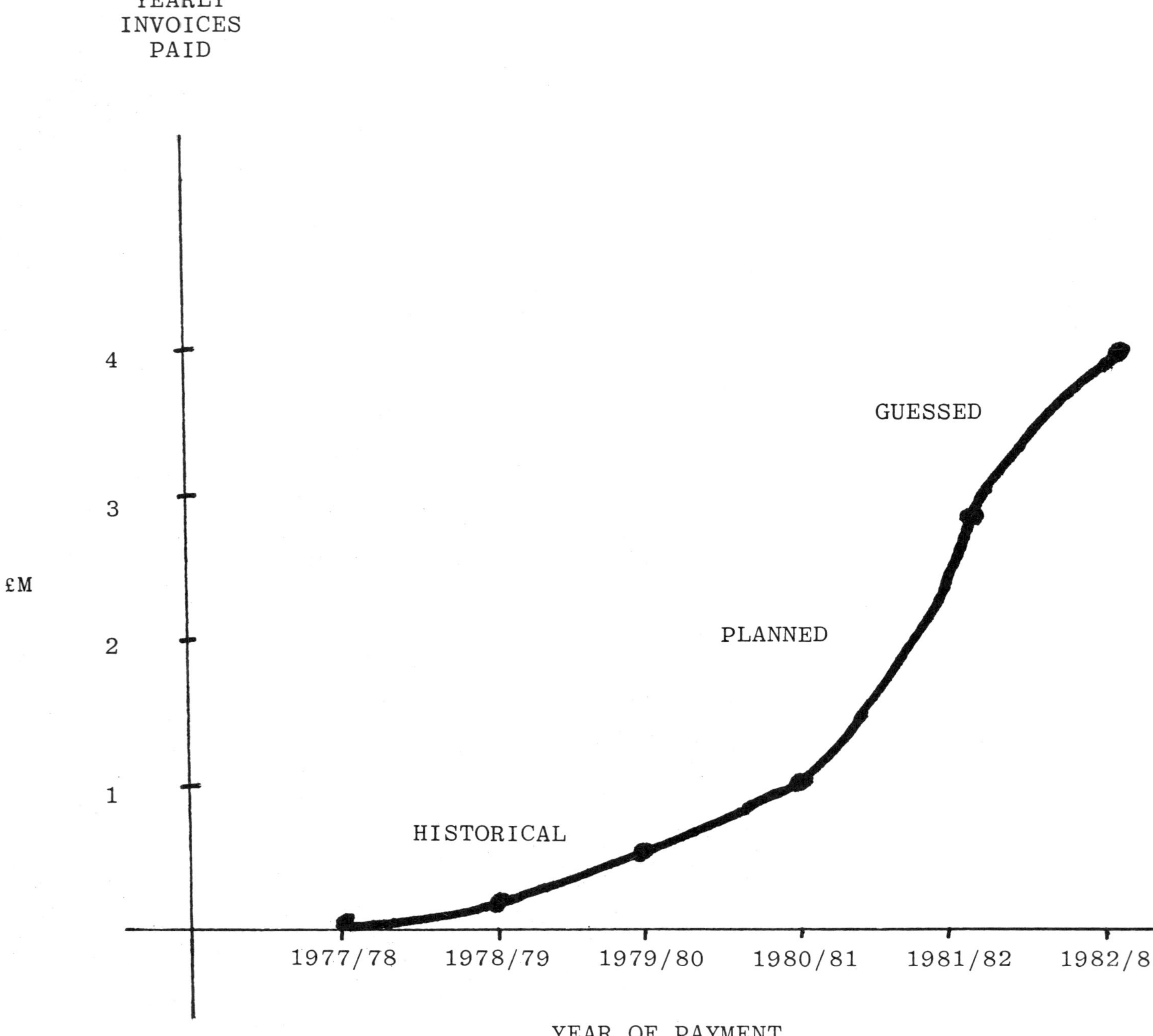

YEARLY INVOICES PAID
£M
4
3
2
1
GUESSED
PLANNED
HISTORICAL
1977/78
1978/79
1979/80
1980/81
1981/82
YEAR OF PAYMENT

THE EUROPEAN WINDPOWER SCENE

S. Hugosson

Stockholm, Sweden
Consultant to Swedyard Corp., Gothenburg

Abstract

Five Western European countries - besides Great Britain - have government financed wind power programmes, aimed at evaluation and/or development of mainly Large Scale wind power units for the production of electricity. The use of small WPU:s is encouraged to some extent. Some other countries have wind power activities, without any real programme. The Soviet Union is developing wind power, and Roumania has recently decided to aim for a certain percentage of wind power in its electricity supply before 1990. This article summarizes the different national activities updated to March 1981.

Introduction

Even though Great Britain undoubtedly is part of Europe, I have understood my task to be to try to cover wind power developments as they go on today in Europe excluding GB. I will do so by describing the scene country by country, and at the end by summing up on time schedules and programme aims. I will only deal with large scale programmes. The pictures I use are not released outside the framework of the IEA agreement on Large Scale WECS development, so they do not accompany this text in printing. I hope GB is soon joining this agreement, to the mutual benefit of improved technical and economical information.

I will first deal with the five countries having government financed wind power programmes, in alphabetical order. I will then give what information there is available concerning activities in some other Western and Eastern European countries.

Denmark

The Danish programme for large electricity producing wind energy systems was initiated in 1977 by the ministry of energy and the electric utilities. The aim is to solve the technical and the technical-economic problems by means of the design, construction and operation of large-scale wind energy conversion systems, and according to this two large machines - the Nibe wind turbines - were erected in 1979.

The two turbines are sited close to each other and are equal, except for the rotor. The main data are:

- 3-blade, upwind rotor, 40 m diameter
- The rotor blades of the A-machine are supported by stays while the blades of the B-machine are selfsupporting.
- The A-machine has 4 discrete pitch angles (start, stop, low wind operation, high wind operation) and is stall-regulated. The B-machine has full pitch control for power regulation
- Blade construction: steel/fibre glass spar, fibre glass shell
- 3-stage gear box, ratio 1:44
- 4-pole asynchronous generator, rated power 630 kW

- Concrete tower, height to rotor hub 45 m
- Wind speeds: cut in 6,5 m/s (at hub height), rated 13-14 m/s, cut-out 25 m/s

The A-machine was connected to the local 20 kV electricity grid for the first time in September 1979 and normal (automatic, unmanned) operation was started in May 1980. The total results as of November 1980 is 845 hours of running and a production of 152 MWh. The result is lower than expected due to some problems with the hydraulics and with the control computer. However, these problems are now overcome. The performance of the A-machine is as expected. The losses in the conversion system (from the rotor to the electricity grid) are however bigger than expected, mainly because of rather big losses (about 40 kW) in the gear-box.

The B-machine was connected to the grid for the first time in August 1980. The commissioning was still going on in November 1980, due to the problems with the control computer. The computer-system of the B-machine must be very quick to do the power regulation by continous control of the blade pitch angle, and it has been necessary to re-program the system to fulfil the requirements.

The construction cost for the two Nibe wind turbines is approximately 15 mill. Dkr. (1980 price level), and the design and development costs are appr. 8 mill. Dkr. Based on the experiences from the design, construction and operation of the turbines, it is estimated to be possible to simplify the machinery corresponding to a cost reduction of nearly 20%, and a further reduction is estimated possibly by mass production and by increasing the dimensions of the system (corresponding to a rotor diameter of 60-80 m). Totally, this should give a construction cost of about 6 000 Dkr/kW.

A "Wind Atlas" has been prepared, a method to calculate the wind speed distribution at a given place and at a given height based on an assessment of the surroundings. The siting possibilities have been calculated based on this investigation of the wind resources and on an assessment of the constraints (conservation interests and other environmental interests, agricultural interests, necessary distances to nearest neighbour for reasons of personal safety, noise and TV/radio-interference etc.). It seems possible to site 1 000 - 2 000 wind turbines - with an annual production equivalent to about 10 per cent of the Danish electricity consumption - on land in Denmark, but not all the machines can be sited on the most windy places, mainly cause to the conservation interests.

The operation and testing of the Nibe machines and the site investigations as well as some studies on aerodynamics and blade materials will continue in 1981. In addition, a preliminary off-shore study will be initiated. The main attention, however, will be given to further development of wind energy systems, where especially the nacelle and the blades will be examined. The nacelle because it counts for nearly half the costs of the Nibe machines, and the blades because they are the only Nibe-components which cannot easily be upgraded. The above mentioned activities will go on in 1981 and the first half of 1982. Thereafter, the continuation of the programme will be discussed based on the results as of mid-1982.

Germany (F.R)

For several years the main German LS WECS activities have been covered by the GROWIAN program. The aims of this program are the development and testing of LS WECS up to the MW scale for electricity production while also assessing the technical, economic and institutional requirements for their use, and stimulating their commercial utilization. In 1980, the GROWIAN program consisted of 25 projects with total costs of about 118 Mill. DM. Besides the development, construction and testing of LS WECS these projects also cover measurements of the wind characteristics for LS WECS, the development of large rotor blades, research for the control of system vibrations and first steps for a profitability analysis.

The project "Construction and test of GROWIAN I" which started in 1979, is making progress and is being carried out under a rather prolonged time schedule. Most of the sub-contracts such as engineering, tower etc., have been placed. The plant will be operated by the "GROWIAN-Bau- und Betriebs-GmbH", a company consisting of the three German utilities HEW, RWE and Schleswag. The location is at the Kaiser-Wilhelm-Koog at the mouth of the river Elbe near the North Sea, where the GROWIAN I is expected to have its "first rotation" in August 1982.

System data of GROWIAN I

Performance

- Rated capacity	3 MW
- Mean annual energy output	12 GWh
- Power-to-area ratio	380 W/m^2

Wind

- Rated wind speed	11,8 m/s
- Cut-in speed	6,3 m/s
- Cut-out speed	24 m/s

Dimensions

- Rotor diameter	100 m
- Rotor speed	18,5 rpm ± 15%
- Hub height above ground	100 m
- Mass of tower head with rotor	240 t

Characteristics

- Rotor blade has welded steel box spar covered with GRP airfoil sections
- Pitch control system operates on the whole blade, and provisions are made for cyclic pitch control

A second, even more advanced LS WECS, called GROWIAN II, is under development. The contractor is Messerschmitt-Bölkow-Blohm GmbH (MBB), Munich. The objective of this project is the investigation of advanced concepts for larger WECS. The construction and testing of a demonstration plant with a scale factor of approx. 1/3 will lead to the preparation of blueprints for a 1/1 prototype plant at the end of 1981. GROWIAN II will have a single blade rotor with horizontal axis and a diameter of 145 m. The hub height will be 120 m. 5 MW power will be produced at a wind speed of 11.5 m/sec. The first rotation of the demonstrator is planned to take place in May 1981 near Bremerhaven.

One 265 kW-Voith-Hütter medium scale WECS is now in construction. Static and dynamic behaviour have been tested in good agreement with theoretical results. "First rotation" is planned to occur at Stötten in summer 1981.

Technical and performance data of Growian II

		1/3 Demo	Growian II
Rotor disc diameter	(m)	48	145
Hub height	(m)	50	120
Power at rated wind speed	(KW_{el})	370	5.000
Power density	(W_{el}/m^2	200	300
Rated wind speed	(m/sec)	10	11,5
Operational wind speed range	(m/sec)	5,7 - 16	6,6 - 18
Annual power production	(GWh/a)	1,3	17
Blade tip speed at rated wind speed	(m/sec)	120	138
Tip speed ratio at rated wind speed		12	12
Maximum overall power coefficient		0,37	0,37
Airfoils		Wortmann FXX77-W-Series	
Max./Min chord length	(m)	2,33/0,56	7/0,68
Twist, root to tip, nonlinear	(deg)	15	15
Rotor blade mass	(kg)	1.400	Appr. 26.000
Nacelle, fully equipped mass	(kg)	28.000	" 300.000

The above-mentioned 25 LS WECS projects are almost entirely funded within the framework of the Programme of Energy Research and Technologies. This programme is part of the activities of the Federal Minister of Research and Technology and to a lesser degree of the Federal Minister for Economic Affairs. The 25 projects amount to total costs of DM 118 Mill. For 1980 the costs of these projects are expected to be DM 40 Mill.; nearly the same sum will occur in 1981.

Some very comprehensive work on wind power integration into existing systems, and the economics of such integration, has been done at the University of Regensburg under Dr. Lorenz Jarass.

Holland

The Dutch government decided to start a national research programme on Wind Energy in February 1976, with the objective to study the technical and economical feasibility of large scale utilization of wind power for the generation of electricity to be fed into the utility grid. The programme has consisted of siting studies, mainly off-shore, research on small

Darrieus turbines, performed by Fokker Aircraft, wind tunnel work on som advanced aerodynamic concepts and on cluster effects, plus the design an construction of one test unit at Petten.

The test unit at Energiecentrum Nederland (ECN) at Petten is a horisontal axis, upwind two-bladed unit of 125 kW power. The rotor blades are fully controlled in pitch. The siting of the unit within a busy research establishment required an elaborate risk analysis, which led to the conclusions - or should we say assumptions - that a catastrophic event with one victim could occur with a frequency of $3x10^{-3}$ a year, while an event with 200 victims - how that could ever occur - could happen once in 10 million years.

Based on a private initiative, a government commission has been studying a scheme, whereby wind power and over-capacity night power would be stored in one or more artificial lakes. These would be situated adjacent to the polders, have a typical area of 300 km^2, with a water level only 16-20 meters above the polder. Installed wind and hydro power would be 2000 MW for one such artificial lake. The Commission should present its findings in November 1980, but nothing is known to me as per March 1981.

Private process industries on the Dutch coast have expressed interest in installing wind power adjacent to the plant, simply to save oil. Their characteristic very high usage of energy, coupled to the modern oil prices, has made them see wind power as a profitable fuel saver option within a near future.

However, there are no decisions as per March 1981 to design or develop or purchase any large scale wind power units in Holland. Neither is anything said about the continuation of the government financed wind power programme after the erection of the test unit at Petten.

Norway

A national research programme in wind energy was started in early 1979. The present plan covers the period 1979 - 1982. The programme was funded with 1,7 million NKR in 1979 and with 2,1 million in 1980, the amount for 1981 not being known at this time.

The programme contains 10 projects, four dealing with wind mapping and wind flow modelling, three dealing with evaluation and specification of wind power units, and three others.

Winds along the Norwegian coast are generally favorable for wind power generation, but the terrain in many places causes trouble through turbulence. There are, however, some fairly flat islands off the coast in some areas, and these are considered as good test locations and possibly also as sites for eventual future groups of turbines.

Norway is almost totally supplied by hydro power for its electrical needs. The grid situation is consequently very well suited for an introduction of wind power. The Norwegian grid is higly integrated with the Swedish, northern Norway is supplied through Sweden. Norway, Sweden, Denmark and Finland maintain strong intertie links and buy and sell eletric power freely between them.

The Norwegian programme is now at a stage where the evaluation of existing units and designs has led up to a decision in principle to purchase one unit of around 600 - 800 kW some time in 1982. This seems to short-cut the original plans with one or two years, by jumping over the small test unit stage.

Sweden

The Swedish programme started in 1975. The primary objective is to investigate the technical and economical possibilities of introducing wind power for large-scale electricity generation. The programme is administered by the National Swedish Board for Energy Source Development (NE). The main part of the programme is related to the design, construction and operation of two large-scale WECS prototypes, located at Maglarp in the province of Skåne in southern Sweden, and Näsudden on the island of Gotland. These projects aim at obtaining a basis for a decision in 1985 about the future of wind power in Sweden. The small experimental unit at Kalkugnen has also given valuable data and experiences in this respect.

An accident occured to this 60 kW experimental wind turbine which had been successfully operated since April 1977. In bringing the nacelle to the ground for a planned change of rotor blades, the mobile crane being used broke down at its yaw bearing. The nacelle and blades fell to the ground and were completely destroyed, together with the crane. An investigation has shown that the accident was caused by overload in combination with insufficient torque applied to the bolted joint keeping the yaw bearing to the crane chassis. A new experimental machine will probably be built, and different alternatives for its localization are being investigated, the most probable one being the Näsudden site (see below).

The contract for design, manufacture and installation of the Maglarp prototype was placed with Karlskronavarvet in July 1979 with Hamilton Standard as a subcontractor. This unit is a two-blades, horistonal axis system having the following main features:

Rated power	3 000 kW
Turbine diameter	78 meters
Hub height	80 meters
Hub type	Teetered
Turbine position	Downwind
Blade material	Filament wound GRP-epoxy
Gearbox	Planetary
Tower material	Steel
Generator type	Synchronous
Cut-in wind speed	6 m/s
Rated wind speed	14 m/s
Cut-out wind speed	21 m/s
Annual output	7-8 GWh

All heavy equipment is on order and hardware manufacture has started. Construction of the nacelle is under way at Karlskronavarvet. Filament winding of the blades has begun at Hamilton Standard. At the Maglarp site, work on the tower foundation has started. Wind measurements in the 120 m meteorological mast have been going on since November 1979, showing slightly better winds than predicted. First rotation of the turbine is expected in November 1981 as originally planned. The machine will be operated by Sydkraft AB, (South Sweden Power Ltd), who are also responsible for site work and project management, under contract with NE.

The contract for the Näsudden prototype was let in September 1979 to

KaMeWa AB with the German firm ERNO as subcontractor. The main characteristics of the two-bladed horizontal axis machine are listed below:

Rated power	2000 kW
Turbine diameter	75 meters
Hub height	80 meters
Hub type	Rigid
Turbine position	Upwind
Blade material	Steel box spar with GRP-epoxy leading and trailing edges
Gearbox	Planetary with bevel stage
Tower material	Concrete
Generator type	Induction
Cut-in wind speed	6 m/s
Rated wind speed	13 m/s
Cut-out wind speed	21 m/s
Annual output	6-7 GWh

Final design review has been completed, leading to an increase in blade thickness and a strengthening of the tower. Welding of the steel blades has begun. At the site, preparations for the tower foundation are under way. The tower will be completed in the fall of 1981. First rotation is planned for June 1982. Wind measurements in a 150 m high mast have been going on since November 1979 showing slightly lower winds than predicted. The Näsudden project is managed by Statens Vattenfallsverk (Swedish State Power Board), who are also responsible for site work and poject management, under contract with NE.

Both Swedish large scale prototypes will be equipped with identical data acquisition systems and with more or less identically located sensors, differences depending on construction of the respective unit. These systems were specified and purchased separately from the prototypes themselves, but the prototype contractors were required from the outset to provide space and supplies for the "brain and nerves" of evaluation, and to accept the responsibility for integrating these systems into their unit, with certain definition deadlines set. This has worked very well.

At both sites, there are information buildings to tell the general public and other visitors basic facts about the wind power generally, the operating utility and the Swedish wind power programme. In the summer of 1981 both information centres will be in operation.

Other Western Europe countries

Two government commissions have been formed in Italy to study different aspects of wind power. The aeronautical industry and the utilities are engaged in these commissions. One large scale concept is to use the daily wind variation across the Italian mountains by locating wind power units in mountain passes, giving considerable potential. No results are known from the deliberations of these commissions.

An experimental unit of about 150 kW size is being designed and built in Spain. A special technological agreement exists, whereby the Spanish industry gets assistance from NASA in the USA. The project was presented at the Wind Power Workshop in Washington in 1979, but its present state is unknown to me, as nothing has been heard at other conferences and as nothing has been published.

Eastern Europe

In the Soviet Union wind power development is managed by a separate institute, very aptly named "Tsiklon" (Cyclone). It is known that experimental units have been built in the Crimea. Units in the range 2-10 kW are manufactured, and some have even been exported, one to Sweden. The smaller units, and probably also medium ones of 25-100 kW, have one use to drive pumps (via electricity) in the irrigation systems on the vast steppes of south-eastern USSR. Articles in the Soviet press indicate that large scale units have been designed and tested, possibly in the Ural mountains.

Following a state visit to Sweden of president Ceaucescu of Roumania, a technical cooperation "umbrella agreement" was signed in 1980. Under that agreement, a technical delegation on wind power has visited Sweden, to study the Swedish plans and to visit industry and utilities. Roumania has a principal decision to cover 5% of its electrical power needs in 1990 by wind power. This means roughly 3 TWh, which corresponds to about 500 large units. It is, however, far too early to say what will come out of all this.

Summing up

In summing up the European situation, I will add some valued comments of my own, and I wish to make quite clear that these comments only reflect my own evaluations.

The Danish programme has been very competently run on a very limited budget, and is aimed straight at useful hardware. There is a certain lack of technological level in the programme, clearly dependent on the small budget, as the necessary know-how is readily available in Denmark. This has given control system problems, and will - presumably - also give structural problems, as load cases and stress calculations could not possibly be thoroughly penetrated within the time-frame and budget given. As Denmark has no other energy sources of its own, wind power is very natural - and traditional - as soon as it is feasible. A decision on "go - no go" can be expected in stages from 1983.

The German programme seems to have too many aims and too many projects. Each new phase in the Growian I subprogramme has meant a more or less fresh start from someone elses blueprints. The blade design for the Growian I is far too complicated. The time schedule has slipped 2-3 years. The Growian II project is extremely far-reaching, and means aiming for the stars before even the roof-top view has been tested. The only practical project is the 265 kW Voith-Hütter unit, based directly on old and tested Hütter designs, but being too small to give really valuable results to use the design of 3000-4000 kW units. I doubt the future of wind power in Germany, and I do not think the Growian I will be up and running before 1983.

The Dutch programme seems to have no defined aim beyond the construction and testing of the Petten unit. The combined wind-hydro schemes now being investigated are interesting, and would seem natural to Holland. Offshore siting of wind power would also come natural to the Dutch, with their vast know-how on structures to be placed in a stormy sea. I presume that the test unit at Petten will be operating this year, but I dare not prophesize about anything after that.

The Norwegian programme is a very rational one, concentrating on mapping the resource and on learning from others experiences. We will very probably see a "semi-large scale" test unit in operation on one of the islands outside Tronheim fiord in 1983. The wind power potential is large enough to be of interest, and system integration would be very easy due to the very high percentage of hydro power in the system.

The Swedish programme is still within its general aims and its general time schedule from 1976, although the prototypes are coming out about one year later than originally proposed in the first programme. This means shorter evaluation, which can be partly offset through the very elaborate testing systems. In late April this year, the Minister of Industry will meet with the power industry and the utilities to discuss an early decision to start planning and designing for one or two small "clusters" of units, probably to be partly financed by government funds. If wind power shows its economy and feasibility, there is a good 50/50 probability that Sweden will start using it in the late 1980-ies.

THE HORIZONTAL AXIS WIND TURBINE PROJECT ON ORKNEY

D. Lindley* and W. Stevenson°
* Taylor Woodrow Construction Ltd, Southall, Middlesex, England.
° North of Scotland Hydro-Electric Board, Edinburgh, Scotland.

Abstract

The paper describes a project to design, construct, commission and monitor two wind turbines on mainland Orkney. The first is 20m diameter and has a rated output of 250 kW, the second is a 60m diameter machine with a rated output of 3MW.

The 3MW machine is based on a two-bladed horizontal axis rotor constructed mainly in steel. The blades will be fixed pitch, except for the outer 20% length of each blade which will be continuously trimmed to provide control of power output. The machine will operate at a nominal 30 revolutions per minute at wind speeds of between 8 and 27 metres per second and deliver the rated output of 3MW at wind speeds between 17 and 27 metres per second.

The 20m diameter machine is similar in most respects but can be operated in both fixed speed and variable speed modes.

Details are given of the site, climatology, anemometry and monitoring programmes as well as economics.

Introduction

A design feasibility and cost study of large wind turbine generators suitable for network connection was undertaken, starting in 1976, by a group comprising British Aerospace Dynamics Group, Cleveland Bridge and Engineering Co. Ltd., Electrical Research Association Ltd, North of Scotland Hydro-Electric Board, South of Scotland Electricity Board and Taylor Woodrow Construction Ltd. The results of this study were summarised in a report titled "Development of Large Wind Turbine Generators (1977)".

This led to a second phase of work in which a reference design was evolved for a 60m diameter turbine. This was rated at 3.7MW at 22 m/s (hub height) and had two fixed pitch steel blades driving an induction generator. It was designed to operate in a strong grid. The study included a comprehensive series of wind tunnel tests to obtain lift and drag data for two dimensional aerofoils at low Reynolds numbers.

A two meter diameter rotating model was also wind tunnel tested to obtain performance measurements. This work was reported in February 1980. In the latter stages of the study however, the situation in the UK with respect to wind generated electricity changed. The North of Scotland Hydro-Electric Board recognised an urgent need to implement wind energy conversion systems on three island systems currently using diesel generated electricity. These are the island groups of Orkney, Shetland and the Western Isles.

The Wind Turbine Demonstration that had long been planned by the UK Department of Energy was therefore presented with the opportunity to produce electricity for customers who needed the power. Electricity consumption is growing in the islands and diesel costs and hence generation costs have been rising sharply. Of these island sites, Orkney was preferred as the site for the first demonstration as a body of wind data was available that suggested the availability of a large number of potential sites that were amongst the best in the UK. Orkney was also the most conveniently located having good access by sea and air. A move to Orkney which has a "weak" electrical grid did, however, mean some changes to the design that had already evolved.

The principle changes include the need for a synchronous generator, as opposed to an induction generator, a "soft" transmission system as opposed to a stiff transmission system and continuous control of rotor torque.

This has led the recently formed Wind Energy Group to design a 20m diameter prototype which models the major features of the 60m machine. This smaller machine will be connected to the grid in a timescale that will allow the 60m design team to benefit in most respects from having designed, assembled, ground tested and commissioned the 20m machine.

Furthermore the diesel system on Orkney represents a set of operational criteria typical of many hundreds of similar systems around the world, thus providing the opportunity to demonstrate machines for a wide international market. The machines still represent a design that can be converted to "strong" UK mainland grids, at high wind speed sites both on-shore and off-shore.

The Island System

The group of islands which comprise the Orkney District have a land area of 97,489 hectares (376.4 sq. miles) and have a population of 18,139 (1979 census). Except for Shetland, the islands of Orkney form the most northerly county of Scotland and at high water there are about 70 separate islands in the archipelago, the main ones of which are shown in Fig. 1.

Kirkwall, the county town of Orkney lies on the largest island called Mainland. The terrain in Mainland to the east and south of Kirkwall is mainly low lying but the terrain to the west and north of the county town is distinctly undulating with numerous low hills rising to 200 to 260 metres. In 1980 the island had 7611 domestic dwellings of which 7211 were occupied. Table 1 indicates the number of consumers on each of the islands supplied, whilst Table 2 gives details of the generating plant providing this supply. Though there has been a constant increase in demand for electricity, the percentage increase has fluctuated due to variation over the years in the severity of the winter weather. Between 1971 and 1978, the annual increase in units sold averaged 11%. Over the last 3 years there has been a marked drop due to the national economy drive on the use of electricity and the self-imposed economy brought on by increased tariffs. The units sold in 1980/81 amount to 74.5×10^6 kWh and on a per consumer basis this amounts to 8854 units per year.

The electrical generators in the Kirkwall power station are driven by large diesel engines detailed in Table 2, and the fuel used is a residual hydro-carbon fuel oil. The engines used are V12 and V16 Hawker Siddeley machines. They generally run for 3000 hours and are then routinely serviced. They have a major overhaul every 2 years or 12000 hours when they are out of commission for 6 weeks. The average life expectancy of the machines is 100,000 hours or 15 years heavy use.

Kirkwall station generates electricity at 11 kV which is transformed at the Power Station to 33 kV for the main distribution system. At main sub-stations it is transformed again to 11 kV to feed the rural networks, being transformed to 415 volts, 3 phase or 240 volts single phase at the consumers premises.

The Sites

Most of the hills on Mainland Orkney were examined to determine their potential as WECS sites. Each was assessed against a criteria which included consideration with respect to road access, proximity to the 33 kV Grid, interference with TV or other communications, height of hill, contours of hill and environmental acceptability.

The sites examined in some detail are numbered 1 to 15 in Figure 2. In summary form, the major factor against the use of a particular site is given below.

Site 1.	Unacceptable due to radio and communication masts already existing on site.
Site 2 and 3	Unacceptable to Local Planning Authority and too far from 33 kV network.
Site 4.	Farm on which site is located up for sale and no chance of early purchase. Too close to cliffs.
Site 5.	Too close to Kirkwall Airport. Possibility of interference with local shipping communications.
Site 6.	Designated a National Scenic Area.
Site 7.	Site of TV transmission mast.
Site 8.	Situated in area subjected to peat cutting rights. Could be long delay in acquiring authority to proceed.
Site 9.	Selected on the recommendation of local Planning Officer.
Site 10 and 11	Low hills. Difficulty in getting a 33 kV overhead line through narrow neck of land to Deerness.
Site 12.	Low hill. Low environmental acceptability.
Site 13.	Poor access. Aesthetically unacceptable.
sites 14 and 15.	Poor access. Pour contours. Aesthetically unacceptable.

Table 3 represents an attempt to rank each of the 15 sites awarding a rating of 0 to 5 for each of the major factors mentioned, and indicates that Burgar Hill, site number 9, ranked the highest on this basis.

Possible interference with communication and transmission paths was a major consideration. Television signals are beamed from Thurso on the Mainland to Keelylang Hill (Site 7) and to the Orkney Island Group from there. Radio signalling together with communication networks (including the Boards) emanate from Wideford Hill (Site 1). A turbine sited on Burgar Hill was considered to have a low probability of interfering with such signals.

The site also had to be big enough for the location of up to 3 large turbines, and Figure 3 shows the location of sites for 3 - 3MW turbines as well as the 250kW turbine. Applications have been made for local planning approval and Section 2 consent and these are now being considered by the appropriate authorities. The Hydro-Board's intentions were made known to the general public via advertising for the first time on 19th March, 1981.

Climatology

The Orkney Islands, with Mainland lying as far north as Stockholm in Sweden, have a remarkable equable climate with no great extremes of heat or cold. Orkney has however a high frequency of gales and strong winds. The generally flat smooth moorland relief of the island, and the almost total absence of woodland means that there is very little natural shelter from the wind.

The major climatological record is that from the Meteorological Office Weather Station at Kirkwall Airport where observations began in 1950. Though other temperature, rainfall and sunshine records are available from elsewhere in the islands, Kirkwall Airport is the only wind recording station in Orkney. This is somewhat unfortunate as there are many locations on Mainland Orkney and elsewhere in the group of islands which are more exposed to the wind than Kirkwall Airport. Plant and Dunsire (1974) also make the point that "in Orkney, many of the hills, knolls and ridges rise very abruptly from the general level of the surrounding terrain and the wind speeds experienced in these exposed upland sites are likely to be considerably higher than at Kirkwall Airport".

Table 4 presents in summary form the frequency of wind speed and direction at Kirkwall Airport which is considered in some detail by Plant et al. Figure 4 gives the annual percentage frequency of winds at Kirkwall.

The highest hourly mean wind speed recorded at Kirkwall up to 1972 was 30 m/s in the hour ending at 1700 hours GMT on 27 January 1961. Up to 1972 the highest gust was recorded at 61 m/s from direction 330 degrees (northwest by north) at 0915 hours GMT on 7 February 1969. At that time it was the highest gust ever recorded at a low level wind recording station in the United Kingdom. Figure 5 plots the monthly variation in mean wind speeds at 10m above the ground from all wind directions at Kirkwall.

The climatological record for Kirkwall available from Meteorological office sources is summarised in Table 5.

Fortunately, the Meteorological office wind speed records at Kirkwall have been supplemented by the data collected by the Electrical Research Association during the 1950's and this work has been reported by Golding (1955), Tagg (1957), Golding and Stodhart (1952). Table 6 below gives details of the ERA sites on Mainland Orkney.

Table 6 ERA Wind Measurement Sites on Mainland Orkney

	Site (1)	Location Nat. Grd. Reference	Altitude Above Sea Level (m)	Height of Instrument Above Grd. (m)	Duration of Records First	Last (2)	Est. Long Long Term Annual Average Wind Speed m/s (3)
1	Costa Hill	N30/311 297	152.4	30/20/36	25.11.48	31.1.52	11.2
2	Vestra Fiold	N30/241 222	131.1	3/20	16.12.48	13.7.49	10.3
3	Bignold Park	N30/454 102	39.6	9	9.7.48	20.2.50	6.6
4	Greenay	N30/296 236	149.0	3	24.6.48	7.8.48	

Footnotes

(i) Sites 1,2 and 4 were considered good potential wind power sites where the estimated annual average wind speed exeeds 8.9 m/s. Measurements were made at Bignold Park for correlation with long term records obtained from a Met-Office station.

(ii) Record not continuous

(iii) At 3m above the ground.

The velocity/duration curve for Costa Hill given by Tagg (1957) has been normalised with the mean wind speed for that site and compared in Figure 6 with data recorded for Kirkwall Airport during the period 1963 - 1972. A Weibull distribution with K = 2.2 and C = 1.15 is seen to be a reasonably good fit to both sets of data.

Elsewhere Golding (1955) has described the work by ERA to investigate the variation of vertical wind gradient with wind direction at the Costa Hill and Vestra Fiold sites. Figure 7 presents results of the hourly wind speed measurements taken at Costa Hill. The curve represents mean values of the wind speed ratio Vh/V66 because the measuring masts commonly used in the ERA survey were 66ft high. Golding also presents data, shown in Figure 8, for vertical wind gradients observed over the same period of time on the summits of both Vestra Fiold and Costa Hill some 7 miles apart and probably subject to identical wind regimes. The hill at Vestra Fiold is less steep than at Costa Hill (Golding gives its average slope as about one in ten as against one in five for Costa) and has resulted in a higher vertical gradient of wind velocity as shown in Table 7.

SITE	RATIOS OF MEAN WIND SPEEDS	
	V35/V10	V66/V10
Costa Hill	1.06	1.11
Vestra Fiold	1.18	1.25

Table 7

The ERA survey also investigated the way in which the lack of symmetry of the hill profile as seen from the different wind directions, influenced the vertical wind gradient. Golding (1955) presents data for ratios of hourly mean wind speed at 66ft (20.2m) to that of 35ft (10.7m) plotted on a polar curve and this is reproduced in Figure 9. He points out that the "readings used in obtaining the ratios for the polar diagram were of wind speeds from directions well distributed round the 360° and included hourly wind speeds up to 60 MPH" (26.8 m/s). The plot shows variations in the ratio V66/V35 between 0.99 and 1.1.

Unfortunately no measurements exist for Burgar Hill though it is likely that it will experience a wind regime not too disimilar from that at Costa Hill and Vestra Fiold. The ERA measurements summarised above do show however the importance of understanding the way in which complex topography affects the vertical gradient of wind speed. Complex terrain will also affect turbulence intensity and other wind structure parameters which in turn have a bearing on wind turbine design, performance and operation. It is for this reason that a comprehensive range of measurements are planned for Burgar Hill and these are described elsewhere in this paper.

On the basis of extrapolations of the Vestra Fiold, Costa Hill and Kirkwall Airport data the long term annual mean wind speed has been estimated for Burgar Hill as 10.5 m/s at 16.3m (hub height of the 20m turbine) and 12.6 m/s at 45m (hub height of the 60m turbine).

The Wind Turbines

The 20m diameter turbine has a rated power of 250kW at 17 m/s and a rotational speed of 88 rpm. It will begin to operate at a wind speed of 8 m/s and shut-down when wind speeds exceed 27 m/s. Based upon the estimate (given in a previous section) that the annual mean wind speed at hub height will be 10.5 m/s, the annual energy output will be 700,000 kWh giving it a capacity factor of 0.32. A more comprehensive description of this machine is given elsewhere in these proceedings by Armstrong, Ketley and Cooper (1981) and of the design of its transmission by Garrad (1981).

The 60m diameter turbine has a rated output of 3MW at 17 m/s and a rotational speed of 30 rpm. Its cut-in and shut-down speeds will be similar to those of the 20m machine. Based on the estimate that the annual mean wind speed at hub height will be 12.6 m/s its annual energy output will be about 10.5 GWh giving it a capacity factor of 0.25.

Both machines are to have a synchronous generator and variable pitch rotor blade tips. It is not possible at this stage to give a more precise specification for the 60m machine as the final design will be subject to modification during the detail design process. Following receipt of contracts it is planned to have the 20m machine commissioned within 12 months. The 60m machine could then be commissioned 2 years after that.

Monitoring and Anemometry

The site layout has been given in Figure 3 and shows the location of the anemometry masts relative to the two machines. Wind measurements are to be carried out for the purposes of site evaluation and design validation and performance assessments of both machines. They will have a long term role together with the machine performance monitoring equipment of providing data for the development of a future generation of WTG designs.

The anemometry work can be considered in three phases. Sometimes running in parallel. These are:-

(i) Site Evaluation

Flow investigations will be made in an atmospheric boundary layer tunnel over a physical model of the site and surrounding terrain. In parallel, short term campaigns using Tethered Aerodynamically Lifting Anemometer (TALA) kites and anemometers on 10m masts will be made at the site itself. These will in turn be supported by measurements of wind speed and direction taken on a 26m mast which will provide short term and long term data sets for correlations with performance measurements made on the 20m diameter machine.

(ii) Wind Data for Performance Measurements

Data will be collected from the 26m high mast for the 20m machine and from sensors on a 80m high mast for the 60m diameter machine. Together this programme will yield data on -

- Mean wind velocity and turbulence profiles and their variation with wind direction and atmospheric stability.
- Power spectral densities of turbulence as a function of height above ground and atmospheric stability.
- Spatial scale and correlation of turbulence.
- Extreme gust probabilities
- Frequency of occurrence of wind speeds for different wind directions.
- Rates of wind veering.

These and other meteorological parameters will be determined during the monitoring of the 60m WTG and correlated with such things as machine performance, turbine blade loads, operation of control tips and power fluctuations.

Monitoring

Both machines are to be monitored to provide data for commissioning trials, validation of design methods and for technology development. A single system shown as a Schematic in Figure 10 is envisaged that will perform all required measurements on both machines.

The measurement objectives can be considered under three headings. Special/Commissioning Measurements, Condition of Interest Measurements, and Long Term Measurements. As an example the long term measurement objectives are listed in abbreviated form in Table 8 together with the parameters that need to be recorded.

Economics

In a twelve-month period of the 1979/80 financial year, 17000 tonnes of heavy fuel oil and 600 tonnes of light oil were used at the Kirkwall power station. The average price of the heavy fuel oil was £86 per metric tonne at that time. The total fuel bill of £1.521 million was therefore equivalent to a fuel cost per unit of 2.002 pence. The average of other generation costs amounted to 0.603 pence so that the total loss in supplying Orkney with electricity amounted to £730,000 for the 1979/80 financial year. The total loss of supplying the Orkney, Shetland and Western Isles amounted to £3.55 million in the same period.

The current cost of heavy fuel oil is £118 per tonne yielding a present cost of fuel per unit generated of 2.8 pence and a total generation cost of about 3.5 pence per kWh.

When the non-recurring cost portion (these include design, dynamic analysis, experimental rigs and testing, and some data acquisition hardware) of the budget cost of £5.6 million for the 60 m diameter turbine is taken into account, the remaining balance of capital costs, and including a construction contingency and project management allowance, is equivalent to a unit cost of approximately 3.2 pence/kWh. This calculation is based on a 15-year life, a 1% O and M charge, 4% fuel cost inflation rate and 5% discount rate. If one assumes a 20-year or 25-year life, the unit cost falls to approximately 2.5 pence and 2.1 pence respectively.

Production versions of this machine could naturally be manufactured, installed and commissioned at a lower cost thereby lowering even further the unit cost of electricity generated.

The Participants

The recently formed Wind Energy Group comprising Taylor Woodrow Construction Ltd, British Aerospace Public Ltd Company and GEC Power Engineering Ltd will design, construct, install and commission both machines for the North of Scotland Hydro-Electric Board (NSHEB). The UK Department of Energy will contribute £4.6 million towards the 60m diameter machine whilst NSHEB will provide the balance of costs amounting to £1 million.

The group, NSHEB and the Department of Industry will fund the design and construction of the 20m diameter turbine. The group will be the main contractor for the anemometry and monitoring phase for both machines which will be funded by the Department of Energy.

Conclusion

The Orkney Project which is still subject to planning consents and contractual arrangements offers a unique opportunity for UK industry and electricity generation authorities to gain operational experience of two of the world's largest wind turbines. The economic, performance and other experimental data will serve the following purposes:

It will provide;

1. Experimental data for the validation of design methods (aerodynamic, performance, aeroelastic, control, electrical etc.).

2. Operational data for refinement of lifetime prediction and economics forecasts and improvement of operational strategy at different penetration ratios.

3. Meteorological and wind structure data sets for the refinement of site selection techniques using physical and numerical models and short term site survey campaigns.

4. A demonstration of a machine design applicable to offshore sites around the UK coast.

5. Credibility for a UK industry to sell worldwide, these and future generations of such machines to electrical authorities operating both stiff and weak electrical grid systems.

Acknowledgements

The authors acknowledge the support given to this project by the UK Department of Energy, the South of Scotland Electricity Board, the North of Scotland Hydro Electric Board, and the Boards of Taylor Woodrow Construction Limited, British Aerospace Public Limited Company and GEC Power Engineering Ltd.

References

Plant, J.A. and Dunsire, A. — 1974, "The Climate of Orkney". Climatological Memorandum No. 71 Meteorological Office, Edinburgh.

Golding, F.W. — 1955 "The Generation of Electricity by Wind Power" published by E. & F.N. Spon Ltd, London.

Tagg, J.R. — 1957 "Wind Data Related to the Generation of Electricity by Wind Power". ERA Technical Report C/T115.

Golding, E.W. and Stodhart, A.H. — 1952 "The Selection and Characterstics of Wind Power Sites". ERA Technical Report C/T108.

— 1979 "Development of Large Wind Turbine Generators". Report WPG 79/3, March UK Department of Energy, London.

Table 1. HYDRO-ELECTRIC - ORKNEY DISTRICT

NUMBER OF CONSUMERS - FEBRUARY 1981

MAINLAND AREA	
Kirkwall	3077
Stromness	1230
West Mainland	1602
East Mainland	1081
	6990
OUTER ISLANDS	
Shapinsay	160
Stronsay	179
Sanday	229
Rousay	118
Wyre	15
Hoy	215
Flotta	85
Graemsay	14
Eday	72
Westray	270
Papa Westray	45
Egilsay	22
	1424
Total	8414

Table 2.

PLANT DETAILS KIRKWALL & FLOTTA
(All sets at Kirkwall except F1 at Flotta)

Set No	Manufacturer	Type	Nameplate Rating MW	CGE Div Rating MW	Installed	Origin	Fuel
1	Mirrlees	KVSS12	2.00	1.6	1956	New	HF
7	Mirrlees	KVSS12	2.00	1.6	1956	New	HF
2	Mirrlees	KVSS16	2.89	2.4	1965	New	HF
8	Mirrlees	MV12 Major	3.52	3.4	1970	New	HF
9	Mirrlees	KV12 Major	3.80	3.6	1972	New	HF
3	Mirrlees	KV12 Major	4.59	4.4	1973	New	HF
6	Mirrlees	MV12 Major	4.59	4.4	1974	New	HF
4	Mirrlees	KV12 Major	4.59	4.4	1976	New	HF
F1	Ruston	TB5000 Gas Turbine	3.275	3.2	1978	New	Gas
5	Mirrlees	KV12 Major	6.3	6.1	1981	New	HF

TOTAL INSTALLED CAPACITY 35.1 MW

Table 3. THE HORIZONTAL WIND TURBINE PROJECT ON ORKNEY - RATING THE POTENTIAL SITES

SITE NO.	1	2	3	4	5	6	7	8	9	10	11	12	13	14	15
A	5	5	5	0	5	5	0	1	3	2	4	4	0	0	0
B	5	0	0	2	5	5	3	3	4	0	0	4	1	5	4
C	0	5	5	5	0	5	0	5	5	5	5	1	5	5	2
D	5	1	4	4	1	0	5	5	4	0	0	0	4	4	5
E	4	0	3	0	2	2	4	5	5	5	5	5	5	2	2
F	0	0	1	2	2	0	0	5	5	5	3	1	0	2	1
TOTAL	19	11	18	13	15	17	12	24	26	17	17	15	15	18	14

KEY A - Road Access
B - Proximity to 33 kV Grid
C - Possibility of Lack of TV or Communication Interference
D - Height of Hill
E - Contours of Hill
F - Evironmental Acceptability

RATING 5 - Good
Reducing to
0 - Bad

TABLE 4

Annual percentage frequency of wind direction and speed at Kirkwall Airport (10 years 1963 to 1972)

Height of vane of Kirkwall anemograph above mean sea level = 41 metres (134 feet)
Height of vane of Kirkwall anemograph above ground = 15 metres (50 feet)
Effective height of Kirkwall anemograph = 10 metres (10 feet)

Hourly mean wind speed	Wind directions in degrees (True)												All Directions
	350-010	020-040	050-070	080-100	110-130	140-160	170-190	200-220	230-250	260-280	290-310	320-340	
0 mph	-	-	-	-	-	-	-	-	-	-	-	-	2.7%
1- 3 "	-	-	-	-	-	-	-	-	-	-	-	-	3.6%
4- 7 "	0.5	0.6	0.6	0.8	0.9	0.8	0.9	0.8	0.7	1.3	0.7	0.5	9.1%
8-12 "	1.2	1.2	1.2	1.5	1.9	2.1	3.2	2.7	1.7	2.1	1.7	1.1	21.6%
13-18 "	1.9	1.5	1.0	1.5	2.8	4.1	5.2	3.4	2.9	3.4	2.5	2.1	32.3%
19-24 "	1.1	0.7	0.3	0.5	1.6	2.2	1.9	1.2	1.8	2.0	1.2	1.1	15.6%
25-31 "	0.7	0.4	0.2	0.4	1.1	1.4	0.7	0.5	1.4	1.6	0.7	0.9	10.0%
32-38 "	0.3	0.1	0.1	0.1	0.4	0.4	0.1	0.1	0.6	0.7	0.2	0.4	3.5%
39-46 "	0.2	0.1	0.0+	0.0+	0.1	0.1	0.0+	0.0+	0.2	0.4	0.1	0.2	1.4%
47-54 "	0.0+	0.0+	0.0+		0.0+	0.0+	0.0+		0.1	0.1	0.0+	0.0+	0.2%
55-63 "	0.0+								0.0+	0.0+	0.0+	0.0+	0.0+%
64-72 "									0.0+	0.0+			0.0+%
> 72 "													0.0%
	5.9	4.6	3.4	4.8	8.8	11.1	12.0	8.7	9.4	11.6	7.1	6.3	100.0%

Notes

1. The above frequencies have been calculated from values of wind direction and speed averaged over each hour during the 10 years from 1963 to 1972.
2. Wind directions are measured in degrees from True North and relate to the direction from which the wind is blowing. For example:

   Direction 360 degrees = wind blowing from North
   Direction 090 degrees = wind blowing from East
   Direction 180 degrees = wind blowing from South
   Direction 270 degrees = wind blowing from West

3. Adding the columns of the above table vertically gives the percentage amount of time in the year with winds from the stated directions.
4. Adding the columns of the above table horizontally gives the percentage amount of time in the year with winds in the stated speed ranges.
5. 0.0+ denotes a frequency of less than 0.05%.

Table 5. Climatological Data For Kirkwall Airport, Orkney

CLIMATOLOGICAL PARAMETER	DESCRIPTION	QUANTITY
Wind Speed at 10m	Hourly mean wind speed (m/s) exceeded 50% of the time.	(1) 6.5
	Once in 50 yr. gust speed (m/s) (3sec gust)	55
	Once in 50 yr. hourly mean wind speed (m/s)	34
	Highest hourly mean wind speed (m/s) recorded to 1974	30
	Highest gust recorded to 1974 (m/s)	61
	Number of days/hours with gusts of 17.4 m/s or more	130/1089
	Number of days/hours with gusts of 24.6 m/s or more	37/193
Snow	Annual mean number of days with snow falling	63.9
	Annual mean number of days with snow lying	15.5
Temperature (°C)	Annual mean daily maximum temperature	10.1
	Annual mean daily minimum temperature	5.3
	Annual mean daily temperature	7.7
	Once in 50 yr. annual maximum temperature	25
	Once in 50 yr. annual minimum temperature	-9.0
	Annual mean number of days with min. temp. $<0^oC$	37.1
Sunshine	Annual average daily duration of bright sunshine (hrs)	3.22
Lightning	Annual average number of days with thunderstorms	<5
Rainfall (mm)	Annual average (depends on exact site and height)	890-1150

Footnote: (1) From Caton (1976).

Table 8 Long Term Measurements

	Objectives	Parameters to be recorded
1.	Determination of electrical power vs wind speed characteristic of the WTG.	power, set power, rotor speed, wind speed and yaw angle.
2.	To quantify the wind climatology over Burgar Hill.	Wind speed and direction, atmospheric stability (meteorological mast).
3.	Determination of availability.	times of contact breaker operation.
4.	To assess the number and frequency of start-up and shut-down operations.	as for item 3 plus wind speed and direction.
5.	Investigate the frequency of tip control operations.	tip angle, wind speed.
6.	Determination of the frequency of occurance of yaw angles and duration. (and orientation operations?)	wind speeds and direction on WTG and meteorological mast.
7.	Assessment of general state of WTG at time of alarm trips or shut-shown trips.	various.
8.	Determination of control system behaviour when confronted with an abnormal operation condition such as a trip or a loss of tip actuation. These records are only long term in the sense that the monitoring system will be scanning for an abnormal condition. Data will only be collected when this occurs.	power demand, grid condition, shaft speeds, shaft torque, tip position, ram position, ram load, gearbox position, control systems snap shut and wind conditions.
9.	Monitoring will allow measurement of the long term performance and reliability of the moving tip system to be evouated. The measurements will be made at intervals, rather than on a daily basis. Any impact of icing should also be detectable.	Tip position, ram position and ram loads, meteorological conditions.

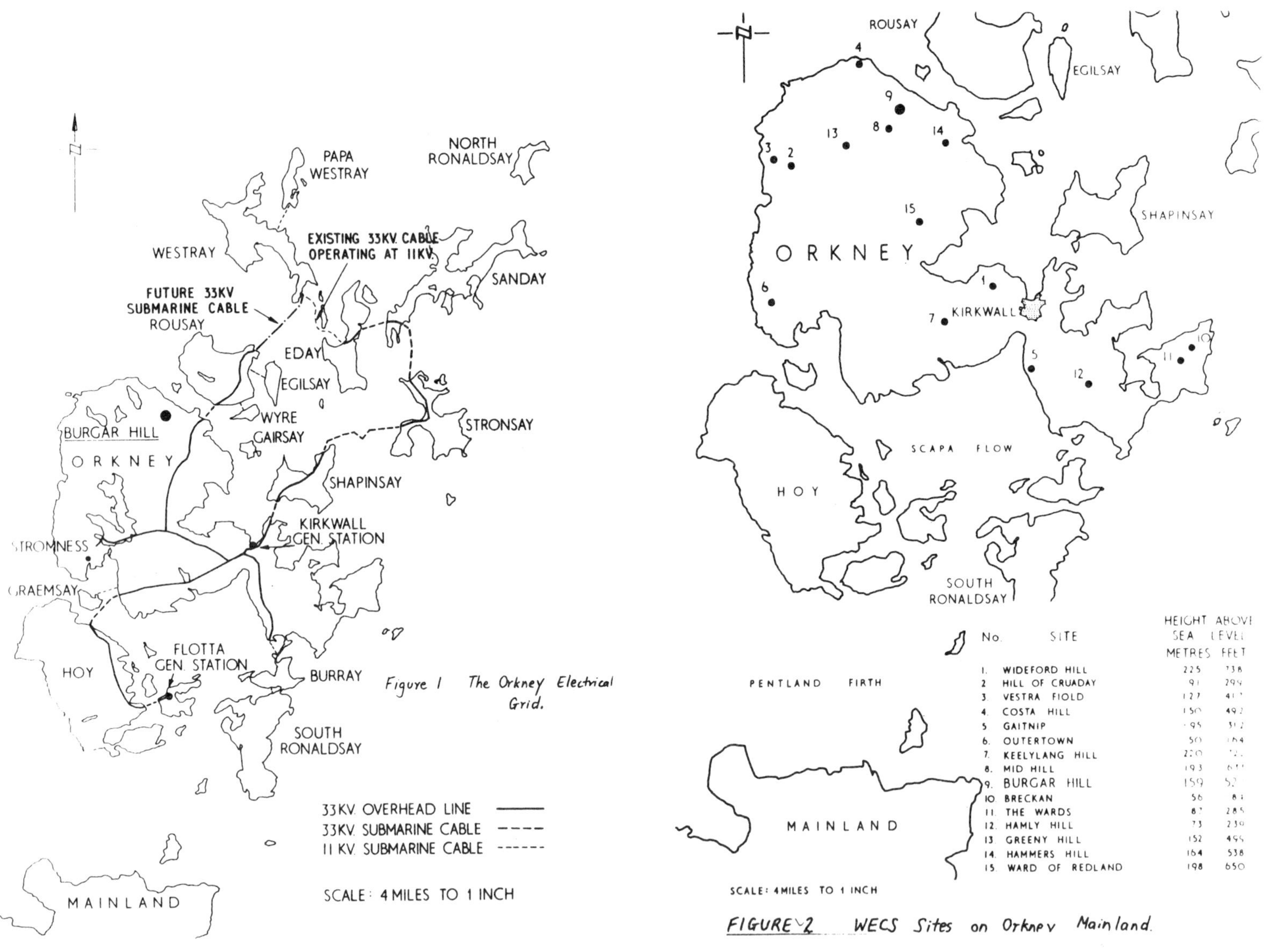

Figure 1 The Orkney Electrical Grid.

No.	SITE	HEIGHT ABOVE SEA LEVEL METRES	FEET
1.	WIDEFORD HILL	225	738
2	HILL OF CRUADAY	91	299
3.	VESTRA FIOLD	127	[illegible]
4.	COSTA HILL	150	492
5	GAITNIP	95	312
6.	OUTERTOWN	50	164
7.	KEELYLANG HILL	220	[illegible]
8.	MID HILL	193	[illegible]
9.	BURGAR HILL	159	[illegible]
10.	BRECKAN	56	[illegible]
11.	THE WARDS	87	285
12.	HAMLY HILL	73	230
13.	GREENY HILL	152	499
14.	HAMMERS HILL	164	538
15.	WARD OF REDLAND	198	650

FIGURE 2 WECS Sites on Orkney Mainland.

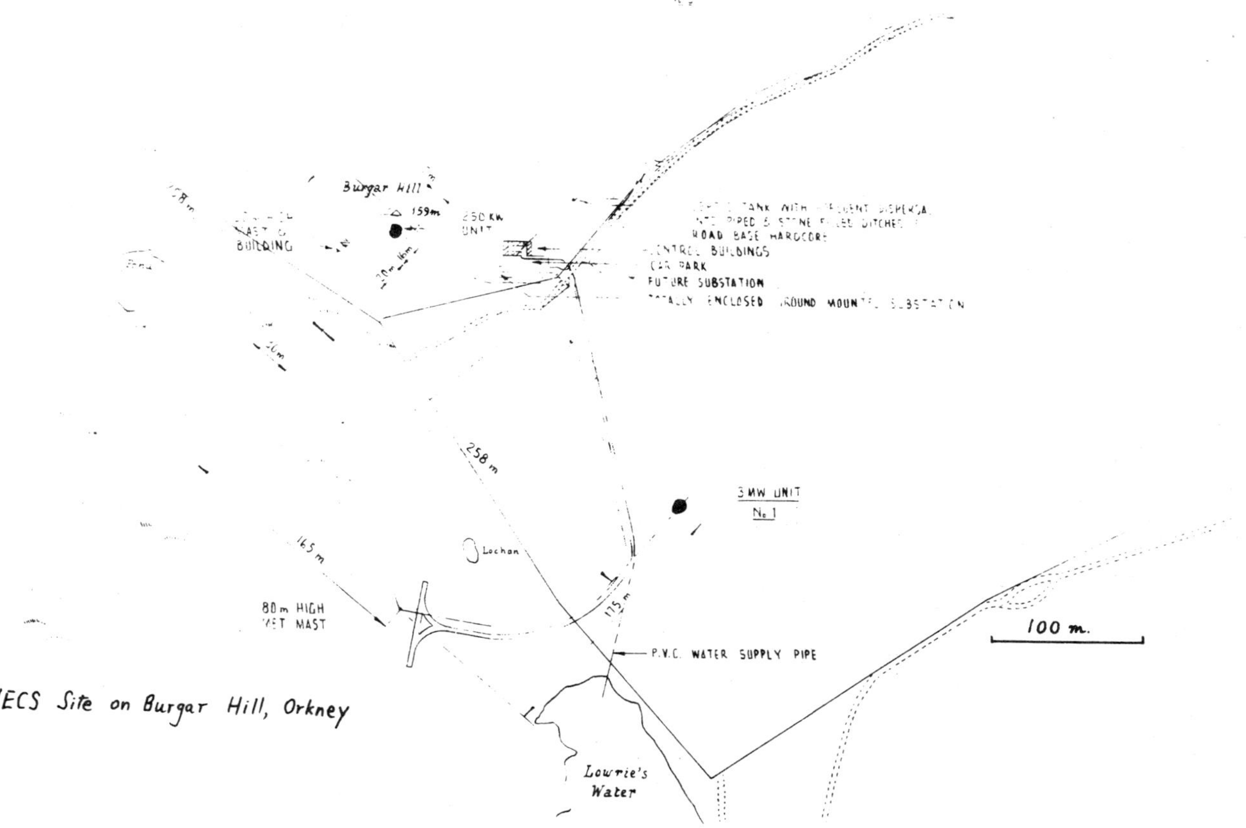

Figure 3 WECS Site on Burgar Hill, Orkney

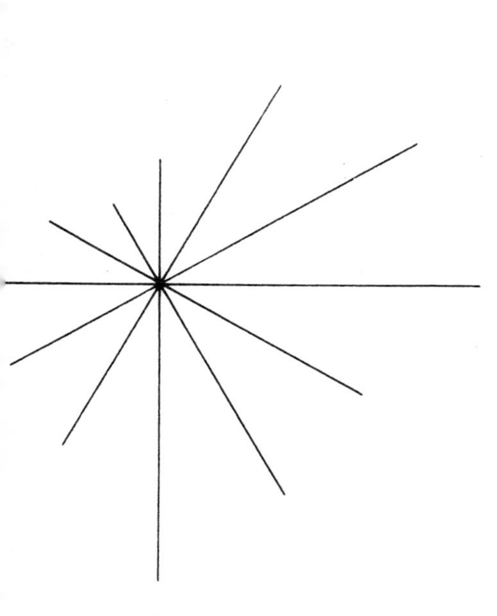

Fig. 4. Annual percentage frequency of winds at Kirkwall Airport (1963 - 1972)

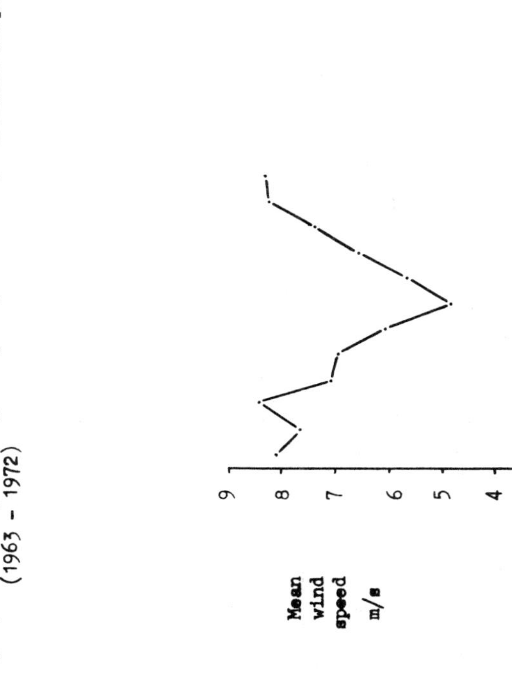

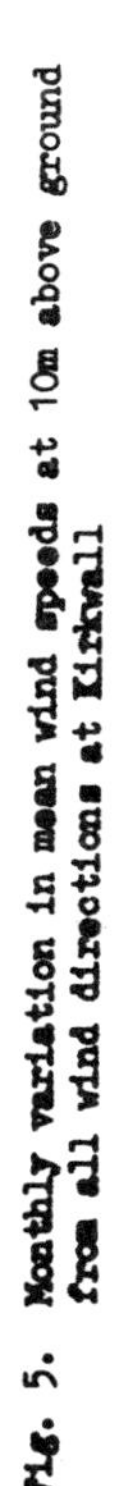

Fig. 5. Monthly variation in mean wind speeds at 10m above ground from all wind directions at Kirkwall

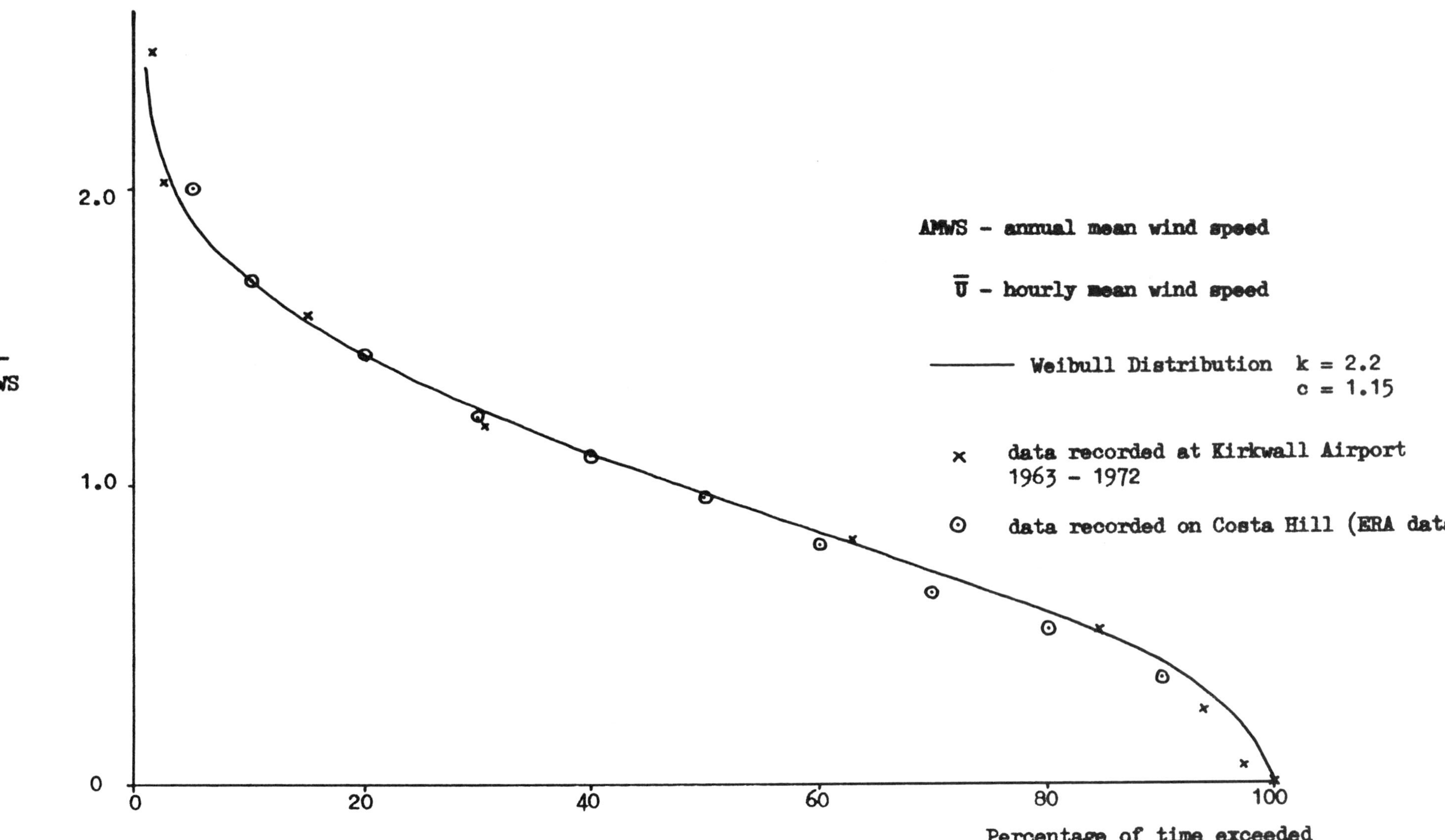

Fig. 6. Normalised Wind Speed Duration Curve

HEIGHT ABOVE GROUND IN FEET
100
80
60
40
20
0
0·6
0·7
0·8
0·9
1·0
1·1
1·2
VH/66 FT

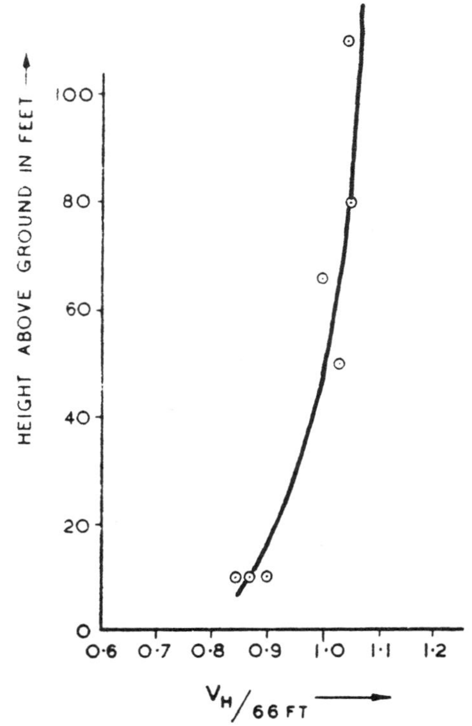

Fig. 7. Variation of wind speed with height Costa Hill

1.2
1.1
1.0
0.9

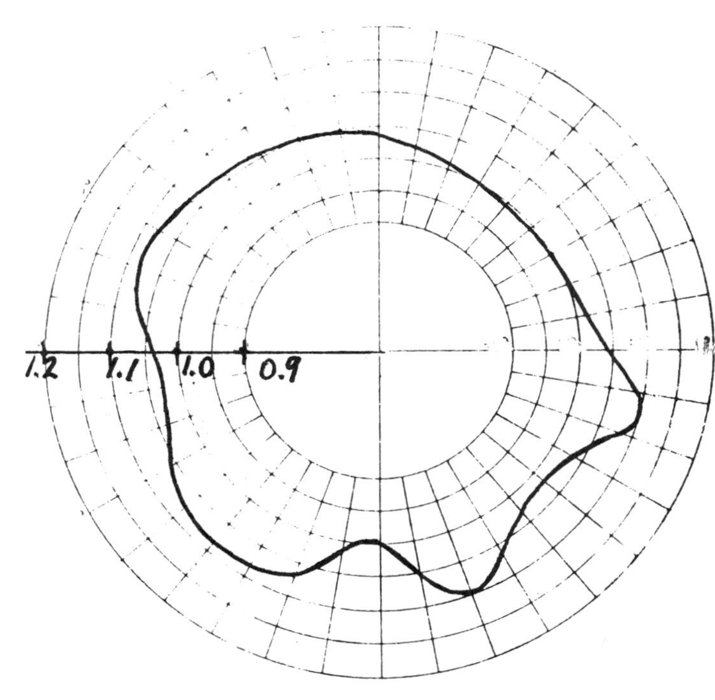

Fig. 9. Polar diagram of vertical wind gradient Costa Hill

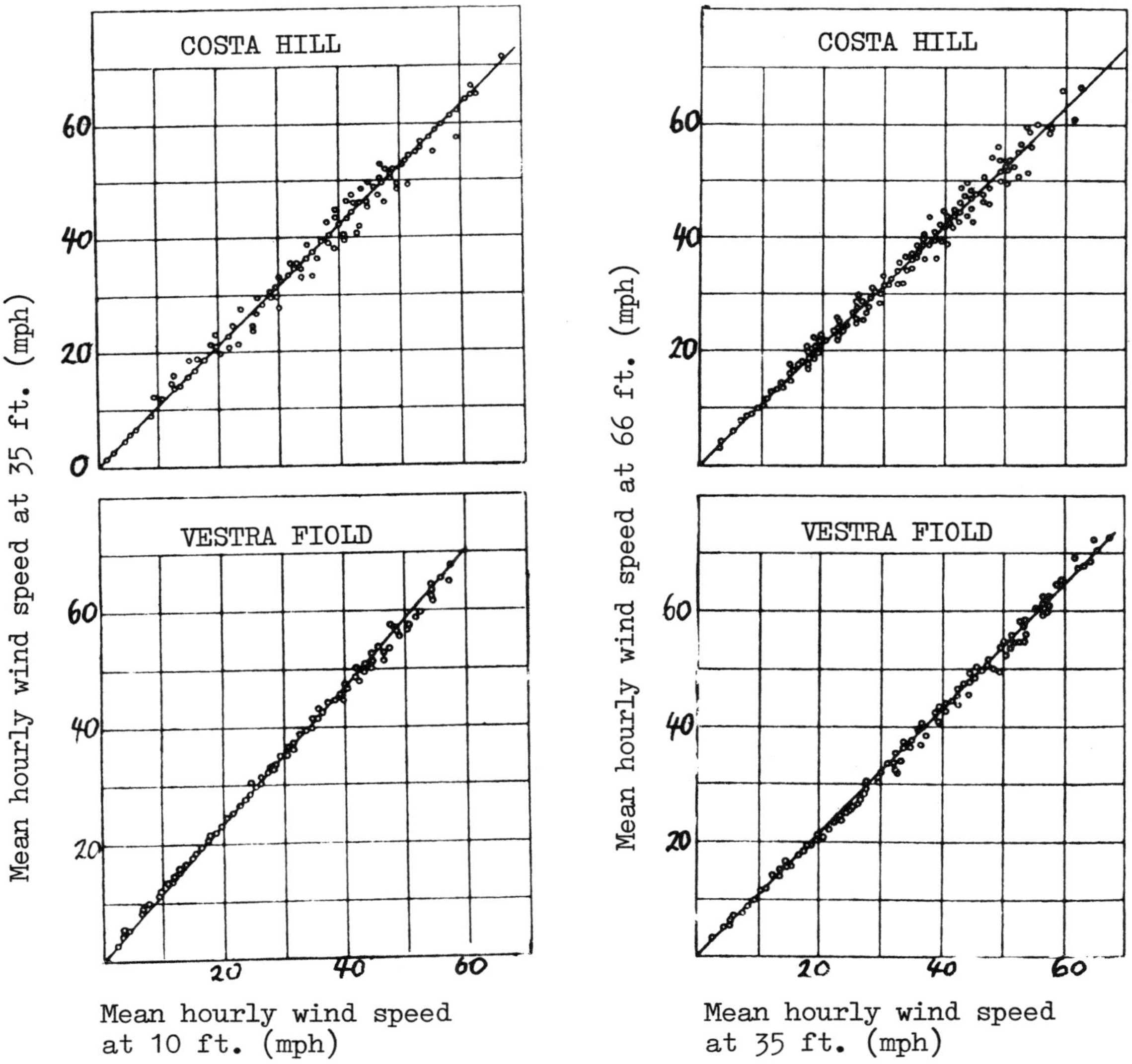

COSTA HILL
COSTA HILL
VESTRA FIOLD
VESTRA FIOLD
Mean hourly wind speed at 35 ft. (mph)
Mean hourly wind speed at 66 ft. (mph)
0
20
40
60
20
40
60
Mean hourly wind speed at 10 ft. (mph)
Mean hourly wind speed at 35 ft. (mph)

Fig. 8. Vertical Wind Gradients

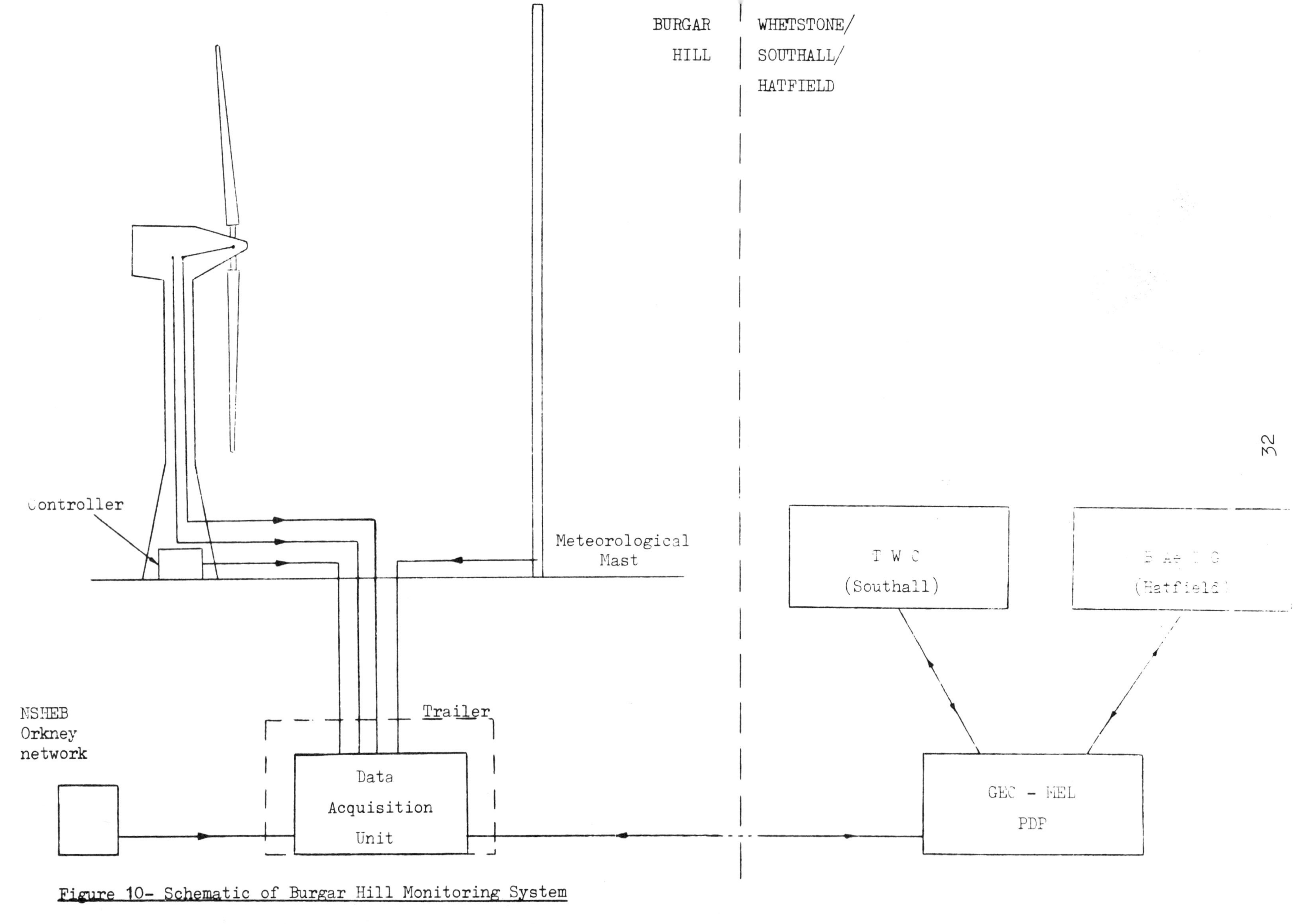

Figure 10- Schematic of Burgar Hill Monitoring System

PROGRESS WITH THE DESIGN OF A 25 METRE DIAMETER VERTICAL AXIS WIND TURBINE

R. Clare, Sir Robert McAlpine & Sons Ltd.,
J. Allan, Aircraft Designs (Bembridge) Ltd.,

INTRODUCTION

The design and working drawings for a 25 metre diameter version of the Musgrove Vertical Axis Wind Turbine are being prepared by an industrial consortium led by Sir Robert McAlpine & Sons Ltd., who will also carry out the civil engineering design. Other members of the Consortium are Aircraft Designs (Bembridge) Ltd., designing the rotor, N.E.I. Clarke Chapman Cranes Ltd., responsible for the mechanical transmission and dynamic analysis of all aspects with Engineering and Power Development Consultants Ltd. designing the electrical and control systems. Dr. Peter Musgrove acts as consultant to the Consortium.

The design study was started on 1st July 1980 and has a planned duration of fifteen months. Support funding is being provided by the Department of Energy. The authors acknowledge permission from the Department of Energy to publish this progress report.

The Vertical Axis Variable Geometry Wind Turbine was developed by Dr. Peter Musgrove at Reading University (1 and 2) from 1975 to 1979. The Department of Energy commissioned a feasibility study of a 25 metre diameter version by British Aerospace, Taylor Woodrow and Reading University and this was reported in July 1979 (3).

The current study is the first stage of a five year plan aimed at the development and construction of an offshore multi megawatt version, although, at present, contractual arrangements with the Department only relate to the present design study. With this ultimate objective in mind the design for the 25 metre diameter version is being prepared, as far as is practicable, to be a scale model of the larger machine. For example, materials for the rotor are being selected as also suitable for the larger machine in an offshore environment. Notwithstanding this prime objective, sight is not being lost of the possible market opportunities for the 130 kw, 25 metre diameter aerogenerator. Such a machine could well be appropriate for duties other than electricity generation feeding into a strong grid and this could result in radical differences in design from those required for the prime objective of the existing study.

ROTOR

At the commencement of the present design study the consortium were concerned about the weight of the blades and the cyclic bending loads in the cross arm that occur when the blades are inclined on the original configuration. Studies were therefore initiated to minimise these problems and the reefing mechanism shown in Fig. 1 was developed.

The reefing mechanism is considerably changed from the original Musgrove design with rigid blades hinged about the cross arm ends. The advantages are :-

(a) Blade bending moments are substantially reduced.
(b) Cyclic bending in the vertical plane on the cross arm when blades are reefed is eliminated.
(c) Actuation jacks are in the plane of the cross arm making the reefing mechanism more compact.
(d) Each blade will be constructed in 2 handed halves, resulting in easier transportation, lower tooling costs and reduced spares holding requirement.
(e) The struts will give a contribution to the power output and are of similar section to the blades.
(f) The torsional stiffness of the rotor is improved.

In modelling the large offshore machine, a maintenance free life of 40 years has been taken as the target for the blades, crossarm and reefing gear. This is important, as engineering appraisal of the structure over the sea and insitu maintenance will be difficult, any repair being possible only at a base workshop at considerable cost. A 40 years life requires the fatigue design to provide for 3×10^8 stress reversals in a marine atmosphere.

With this number of cycles, the maximum stress in aluminium would be 2000 p.s.i. whilst with steel 10,000 p.s.i would be the limit resulting in a heavy rotor requiring sophisticated treatment and coating to prevent corrosion.
With these traditional materials not favoured, the choice rested with titanium or composite materials. This conclusion was in line with the feasibility study, which appeared to favour the use of titanium. However, enquiries have revealed the likelihood of major increases in the price of titanium mainly due to its important strategic role. There are also problems regarding bonding of titanium. Hence it is not proposed to use titanium as the principal material though it has important secondary uses in the rotor.

Carbon fibre reinforced plastic (C.F.R.P.) construction provides for low mass, freedom from corrosion and a nominal fatigue strength approaching ultimate strength, due to the moderate strains induced in the matrix resin. Glass and Kevlar fibres are strong but of much lower stiffness and neither can approach C.F.R.P. fatigue endurance when set in epoxy.

It is important that full advantage is taken of the use of C.F.R.P materials. Not only is the fatigue life immensely improved but the need for maintenance and inspection is virtually eliminated. This reduced inspection may be difficult for some authorities to accept but it must be realised that no signs of wear or fatigue will be visible with C.F.R.P. and all other components will be either in titanium or designed at very low stresses to give long lives. Nor will any ultimate failure be catastrophic; there will be adequate warning of deterioration from the monitoring of vibration and performance, any deterioration will automatically result in brake application. It is unlikely that a blade could fly off. In other words the windmill could be run until it was stopped automatically. All components requiring regular maintenance will be within the tower and mostly at ground level. The only exception to this will be the actuating jacks of the furling mechanism and access to these will be possible.

At present carbon fibre is a relatively expensive material due principally to the limited demand. It is confidently anticipated that demand will substantially increase and result in significant price reductions.

It is proposed that the blades will be in carbon sheet bonded to a Nomex honeycomb core, to form covers to the spars and ribs which will be in carbon or glass fibre as appropriate. Titanium will be used to form leading and trailing edges, also for self locking connecting bolts. Fig. 2 shows a typical cross section of the blades for the 25 metre diameter rotor.

The carbon fibre will provide an adequate path for lightning conduction.

Stresses in the C.F.R.P. will be in the order of 25,000 p.s.i at rated power level. Shear will be transmitted by interlocking shear strips on spars and ribs, not by bolts. Bond stresses will be limited to 250 p.s.i. and bearing stresses to 10,000 p.s.i. The shear strips are a novel design and they will be subjected to fatigue testing during the design programme.

Loadings on the crossarm are not major and design of this member is proceeding in mild steel suitably protected with a flame applied coating of aluminium.

Actuation of the reefing will be achieved hydraulically and in operation, the blades will vary from the normal vertical position to fully reefed at 70° within one minute. Hydraulic accumulators will be provided which will permit faster reefing in emergency situations. Reefing will be controlled by sensing the torque and a delay will be built into the electronic system to enable changes in torque to be sensed over 60 seconds prior to change in reefing angle Normal rotor operation will be at 27 r.p.m.

MECHANICAL TRANSMISSION

Hydraulic transmission has been considered but is not proposed for the 25m machine. The main gearbox will be at ground level but it is proposed to have the first speed step up at the top of the tower, beneath the rotor bearing, in order to reduce the size of the transmission shaft. The design will make adequate allowance for surge torques and emergency out of balance loads. Different transmission options along this theme are currently being considered.

A multiple braking system will be provided.

ELECTRICAL SYSTEM

The principal design is intended for linking into a strong grid and a generator will be provided that can operate in a synchronous or induction mode with a capacity of 160 kW and running at 1500 r.p.m. This will give the opportunity of testing with both types of generator.

All systems will be electrically controlled including the hydraulic blade actuation. The prototype will be extensively instrumented and monitored.

The electrical design, as with the rotor, will also consider application of the windmill to duties other than linking into a strong grid when preference may be for different types of generator.

TOWER AND FOUNDATION

Alternative designs in steel and concrete are being considered for the tower but in either case a rigid tower is favoured for the 25 metre prototype. A tubular tower is regarded as more aesthetically pleasing than an open lattice structure and in addition gives a high level of protection to the mechanical and electrical transmission. A tower diameter of about 2.7m looks probable.

The alternatives of raft and piled foundation will be reviewed to cater for a variety of foundation conditions. The dynamic analysis is likely to be the major controlling factor in the design of foundation and tower.

MODEL EXPERIMENTATION

Closed circuit and open jet wind tunnel tests are both included in the current phase as well as the construction of a 6 metre diameter free-air model. The programme of wind tunnel tests is underway.

It is proposed to erect the 6 metre model at Bembridge, Isle of Wight. This model will have relatively sophisticated controls and it is anticipated that the following experimental benefits will be derived from its performance.

1. Full modelling of the automatic reefing mechanism with partial modelling of the hydraulics.
2. Improvements in fairing shapes by visual flow methods.
3. Experimentation with generator characteristics and electrical control system.
4. Response to gusts by torque meter readout.
5. Start characteristics at various reef angles.
6. With a constant speed generator the Cp may be gauged continuously from the torque meter readout.
7. Experience of power readings over a considerable range of wind speeds and furling angles.

Limitations on interpretation of results will be the Reynolds number effects but these will be less severe than on the 3m diameter wind tunnel model. Considerable wind turbulence will also be experienced at the selected hill top site at Bembridge thus giving an effective test of the system under an arduous wind regime. The hill site is close to the coast with an elevation of 350 ft. OD.

The model has laminated spruce blades with glass fibre coating and the tower will be a modification of the standard tower and transmission system of the 6 metre diameter wind turbine produced by P.I. Specialist Engineers Limited.

CONCLUSION

The design of the 25 metre diameter vertical axis aerogenerator will be sufficiently advanced for building to commence in autumn of this year.

REFERENCES

1. P.J. Musgrove. "Variable Geometry Vertical Axis Windmill". First International Symposium on Wind Energy Systems 1976.

2. P.J. Musgrove and I.D. Mays. "Development of the Variable Geometry Vertical Axis Windmill". Second International Symposium on Wind Energy Systems, 1978.

3. British Aerospace Ltd., Taylor Woodrow Construction Ltd., Reading University. Feasibility Study, 25 metre Variable Geometry Vertical Axis Windmill. Report to Department of Energy.

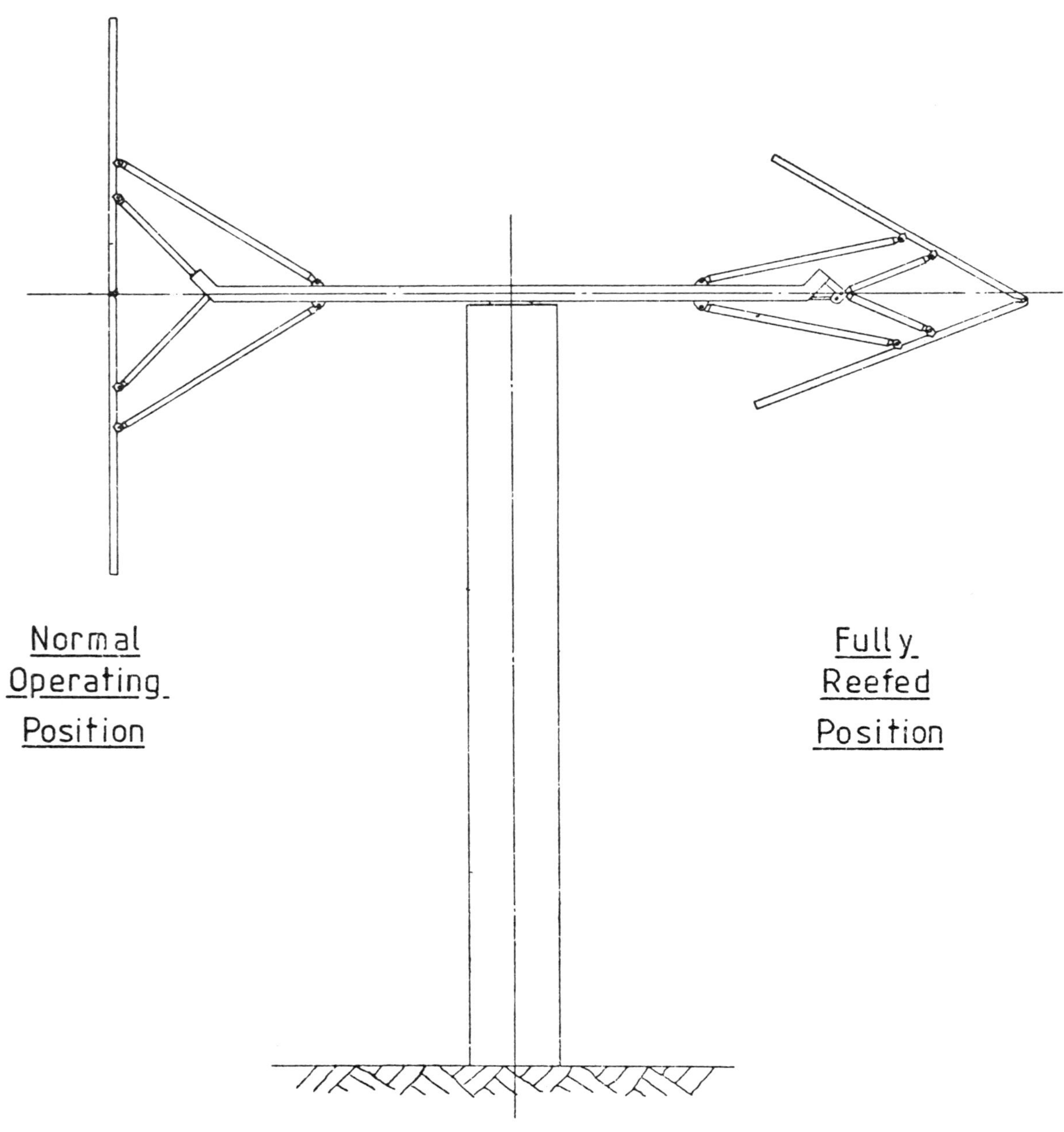

VERTICAL AXIS WIND TURBINE DIAGRAM SHOWING ROTOR REEFING MECHANISM

FIG. 1.

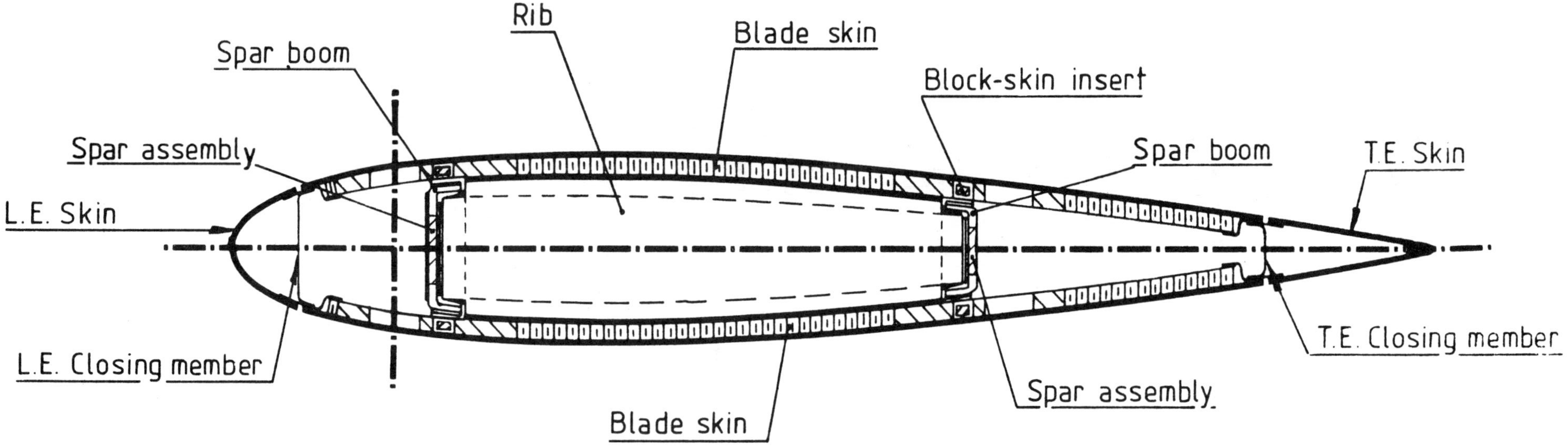

SECTION OF BLADE

25m Dia. VERTICAL AXIS WIND TURBINE

FIG. 2.

WIND TURBINE TRANSMISSION SYSTEMS

A.D. Garrad

Taylor Woodrow Construction
Southall, Middlesex

Abstract

Different types of soft transmission systems are briefly reviewed. The "swinging gearbox" transmission is then examined using both direct integration and frequency domain techniques. Shortcomings in tools available for this type of analysis are discussed.

Introduction

The dynamic characteristics of the transmission system of a wind turbine play a very important role in determining the quality of power that it produces.

Electrical grids may be described in terms of their strength - a strong grid is one which can accommodate fluctuations in input power without adverse effect; a weak grid does not have this ability, and a wind turbine supplying such a grid must be capable of attenuating the torque spikes produced by gusts of wind. The use of a "soft" transmission system is the most common way of achieving this objective. Designing softness into a dynamic system increases the likelihood of dynamic instabilities and therefore careful analysis is essential.

One of the major problems with the analysis of wind turbines in general is the stochastic nature of the wind induced loads. The discrete gust and the spectral approaches are both in popular use for characterising this type of loading.

There are drawbacks with both of these methods. Discrete gusts are useful in order to visualise the time domain behaviour of a system, but realistic representation of a discrete gust is very difficult. For buildings and fixed wing aircraft the spectral approach is considerably more satisfactory and realistic spectra have long been established. Such a firm basis does not, however, exist for the loading on a rotating blade - the way in which the rotational sampling of a blade transforms the wind spectrum is complex and has so far received little attention.

Despite the shortcomings of both the time and frequency domain approaches, they are all that exist at present and in the analysis that follows both will be adopted.

References will not be refered to individually but a list of useful papers is included at the end of the text.

Soft Transmissions

In a soft transmission attempts are made to introduce extra flexibility into the system. This can be done by several means, some of which are listed below:

- i) quill shaft
- ii) fluid coupling
- iii) flexibly mounted gearbox

A quill shaft was adopted for Mod 2. A typical power spike for this machine is about 10%. This form of soft transmission is the simplest available. It does, however, have several inherent problems. The shaft is an expensive item and if the transmission is found to perform unsatisfactorally it will be difficult to rectify. In addition, a steel shaft has little structural damping, so there is a possibility of dynamic problems arising from the large inertia of the rotor oscillating at the end of a long, soft, undamped shaft.

A fluid coupling device allows slip to occur between its input and output. This permits very good quality power to be obtained, but there are accompanying penalties. For 2% slip on a 250 kW machine 5 kW of heat would be generated within the nacelle. This is problem enough for a small machine but becomes prohibitive when extended to large megawatt size machines. Mod O has a fluid coupling which is now working satisfactorally but it has created considerable problems in the past.

The remaining item in the list of possibilities is the flexibly mounted gearbox. This approach, which is commonplace in ship transmissions, has been adopted by Hamilton Standard on WTS 4, GE on Mod 5 and by the Wind Energy Group on the Orkney 20m machine. This method offers several advantages over other solutions. The ability to insert an external damper with relative ease allows considerable control over the behaviour of the system. The spring rate may be chosen to locate the eigenfrequencies where they are required. If the load spectrum is not quite as anticipated during the machine design it should be easy to make the necessary adjustments.

The philosophy behind this design is that when a gust of wind is incident on the rotor disc, the resulting torque spike transmitted through the drive train causes the angular displacement of the gearcase to change. Some of the energy contained in the spike is thus absorbed rather than transmitted to the generator and grid.

Analysis of the Orkney 20m Transmission

A schematic drawing of the Orkney 20m transmission system, together with the nomenclature used in the following analysis is shown in Figure 1.

In addition to the symbols shown there, θ is used to denote the rotation of each inertia e.g. θ_1 is the rotation of the rotor. ϕ denotes the displacement of the gearcase - W_s is the synchronous speed, n_1 and n_2 are the gear ratios of the first and second stages respectively, and the gear ratio of the whole gearbox is $n_1 n_2 = n$.

The dynamics of the synchronous generator itself are very complex. The intention of this paper is to outline a simple analysis of the drive train and so a suitably simple generator model will be adopted. A reasonable representation of the electrical components is to consider the alternator inertia connected by a parallel spring-damper system (K_e, C_e) to an infinite inertia rotating at W_s which represents the grid. K_e and C_e are frequency and power dependant. The power dependence will be neglected here and C_e and K_e are represented by:

$$C_e = C_{e_o} + Af$$

$$K_e = K_{e_o} - Bf$$

Where f is the frequency at which the power train is excited.

From kinematic consideration of the gearbox (1)

$$\theta_5 = -L\phi + n\theta_2$$

Where L is a linear operator

The shafts within the gearbox are assumed stiff, so

$$\theta_3 = \theta_4$$

From dynamic consideration of the gearbox simplified equations of motion may be derived:

$$(J' + \Gamma_2)\ddot{\phi} + C\dot{\phi} + K\phi = (L/n)\,T_{12} + \Gamma_1\ddot{\theta}_2$$
$$\Gamma_4\ddot{\phi} = n\,T_{65} - T_{12} + \Gamma_3\ddot{\theta}_2 \qquad (2)$$

Where $\Gamma_1, \Gamma_2, \Gamma_3, \Gamma_4$ and J' are inertias derived from I_1, I_2, I_3, I_4 and J_1.

T_{ij} is the torque applied to inertia j by the shaft joining inertia i to inertia j. If T_a is the aerodynamic input torque, the remaining equations of motion may be written:

$$T_a - K_1(\theta_1 - \theta_2) = I_1\ddot{\theta}_1$$
$$T_{12} = K_1(\theta_1 - \theta_2) + C_1(\dot{\theta}_1 - \dot{\theta}_2)$$
$$T_{65} = K_5(\theta_6 - \theta_5) + C_5(\dot{\theta}_6 - \dot{\theta}_5) \qquad (3)$$
$$T_e = K_e(\theta_6 - \theta_7) + C_e(\dot{\theta}_6 - \dot{\theta}_7)$$
$$\theta_7 = W_s t \text{ and } \dot{\theta}_7 = W_s$$

Only perturbations from the equilibrium position will be of interest, in which case W_s can be set to zero without loss of generality.

Methods of Analysis

Two distinct methods were used to analyse the system. The complete set of equations of motion which describe the system shown in Figure 1 are given above.

These were first solved numerically in the time domain. This was time consuming and expensive. A simple static analysis of the power train showed that the effective stiffness of the system was dominated by the gearbox mounting spring K and the electrical "spring" K_e. It was thought reasonable in the light of this to reduce the complexity of the problem by performing a "stiff shaft" analysis i.e. putting $K_1 = K_5 = \infty$. A comparison of time domain solutions for the simplified and full sets of equations showed that the stiff shaft analysis was indeed a good approximation to the real system. This simplification, together with neglecting the inertia of the gearwheels I_2, I_3, I_4 and I_5, which are small in comparison to I_1 and I_6 when referred to the same shaft produces a much more tractable set of equations:

$$J\ddot{\phi} + c\dot{\phi} + K\phi = (-L/n)\,T_1$$
$$T_a - T_1 = I_1\ddot{\theta}_1$$
$$T_1 = n(I_6\ddot{\theta}_5 + K_e(\theta_5 - \theta_7) + C_e(\dot{\theta}_5 - \dot{\theta}_7)) \qquad (4)$$
$$\theta_5 = n\theta_1 - L\phi$$

Time Domain Analysis

It is probably easier to appreciate the behaviour of the transmission system by considering its behaviour in the time domain. To this end Figure 2 shows the response to both high frequency random fluctuations and to a square wave input. The first eigenfrequency of the system is about 0.5 Hz and so the transmission only works effectively as a filter for somewhat higher frequency distrubances. This behaviour is exemplified in Figure 2. The filter is quite efficient for the "turbulent wind" type input, but only serves to round off the edges of the square wave gust. Reference to the square wave input indicates that the damping of the system is slightly under critical. The system whose response is shown in Figure 2 has not been optimised.

It is clumsy to conduct parametric studies in the time domain, and for this reason a frequency domain approach was adopted for most of the analysis.

Deriviation of Frequency Response of Stiff Shift System

By setting $W_s = 0$ in equations 2 and 3, they can be reduced to a pair of simultaneous differential equations:

$$(J + (L/n)^2 I_1)\ddot{\phi} + c\dot{\phi} + K\phi - (L/n^2) I_1 \ddot{\theta}_5 = -L/n \, T_a$$
$$(I_6 + I_1/n^2)\ddot{\theta}_5 + C_e\dot{\theta}_5 + K_e\theta_5 - (L/n^2) I_1 \ddot{\phi} = 1/n \, T_a \qquad (5)$$

Assuming
$$T_a = T_a^o e^{iwt}$$
$$\theta_5 = \theta_o e^{i(wt+\xi)}$$
$$\phi = \phi_o e^{i(wt+\psi)}$$

These equations reduce to:

$$\begin{bmatrix} H_1 e^{i\xi} & H_2 e^{i\psi} \\ L_1 e^{i\xi} & L_2 e^{i\psi} \end{bmatrix} \begin{bmatrix} \theta_o \\ \phi_o \end{bmatrix} = \begin{bmatrix} T_o/n \\ (-L/n) T_o \end{bmatrix} \qquad (6)$$

$$\text{or} \quad \underset{\sim}{M} \quad \underset{\sim}{\theta} \doteq \underset{\sim}{T}$$

If
$$\Delta = det(\underline{M})$$
$$= (H_1 L_2 - H_2 L_1) e^{i(\xi+\psi)}$$

Then
$$\theta_o = (T_a^o / n\Delta)(L_2 + LH_2) e^{i\psi} \qquad (7)$$
$$= (T_a^o/n) \left[\frac{L_2 + LH_2}{H_1 L_2 - H_2 L_1}\right] e^{-i\xi}$$

We require an expression for T_e. Put $T_e = T_e^o e^{i(wt+\mu)}$, using equation (7) this reduces to

$$T_e^o e^{i\mu} = (K_e + i\omega C_e)\,\theta_o e^{i\xi}$$

$$\text{so } T_e^o = (K_e + i\omega C_e) \frac{L_2 + LH_2}{H_1 L_2 - H_2 L_1} \frac{1}{n} T_a^o e^{i(\xi-\mu)} \qquad (8)$$

$$= \frac{H(\omega)}{n} T_a^o$$

The frequency response relating the aerodynamic and electrical torques has therefore been established.

Discussion of Frequency Response

$|H|$ is the system gain. The ideal shape for $|H|$ would be a sharp step from 1 to some small value at a very low frequency. Such behaviour would indicate that the transmission would pass very low frequency oscillations, but filter out any that occur at higher frequencies.

Figure 3 shows how $|H|$ is effected by changes in K and C. K is largely fixed for the reasons explained above, and within its possible range of values changes do not have an important effect. C is by far the most influential parameter. Judicious choice of its value is important since, for a given value of J and K, it determines the height of the peak in H and the sharpness of "roll-off" in the characteristic.

A typical curve has a peak value of $|H|$ = 1.2 and a reasonably sharp cut-off; the attenuation is 0.25 at 2Hz. This means that fluctuations of 60% in the input torque will be reduced to 15% at the generator for frequencies above 2Hz. The shape of the curve is relatively insensitive to the value of J.

Figure 3 also shows the "control" case when the gearbox is rigidly mounted. $|H|$ is plotted both with and without the frequency dependence of C_e and K_e. The difference between these two curves is very marked because K_e is the only flexibility in the system. In fact, the real control case will be somewhere between these two curves. The lower one can be used to make conservative estimates of how well the soft transmission is behaving.

The shortcomings in the spectra available as input to be transfer function have already been discussed. It is, however, instructive to look at the input and output spectra of the system. The von Karman spectrum was chosen as the most realistic model and was used as input to the system transfer function. This spectrum and the resulting output spectrum+ are shown in Figure 4. It is only intended that the gearbox should filter out perturbations of frequencies higher than about 1.5 - 2 Hz, lower frequencies will be removed by other means. With this in mind Figure 4 shows that the soft transmission is doing its job quite well.

In addition to the von Karman spectrum in Figure 4, there is also a sharp peak at a frequency of 2P (twice per revolution). This is an approximate representation of how the shear layer will modify the spectrum. $|H|^2$ at 2P is about 6% and so it has quite a dramatic effect on the peak.

Concluding Remarks

Some simple methods for analysis of a fairly complex branched transmission system have been outlined. Although some numerical methods have been employed, emphasis has been placed on analytical methods which produce adequate solutions without recourse to computers.

The behaviour of a soft transmission with a swinging gearbox has been discussed and the method shown to produce desirable characteristics.

It is felt that the absence of a proper spectrum for the analysis of rotating systems represents an important gap in the tools available for this type of work.

+ $S_{out}(f) = |H(\omega)|^2 S_{in}(f)$

References

1. Dynamics of Wind Generators on Electric Utility Networks
   C.C. Johnson and R.T. Smith
   IEEE Trans. AES Vol. AES 12, No. 4 July 1976

2. Drive Train Normal Modes Analysis for ERDA/NASA 100-KW Wind Turbine Generator.
   T.L. Sullivan, D.R. Miller and D.A. Spera
   NASA TM-73718 July 1977

3. Power Train Analysis for the DoE/NASA 100KW Wind Turbine Generator
   R.C. Seidel, H. Gold and L.M. Wenzel
   NASA TM-78997 October 1978

4. On Wind Turbine Power Measurements
   S. Frandsen and C.J. Christensen
   Proc. 3rd International Symposium on Wind Energy Systems, Copenhagen August 1980.

5. Wind Energy Conversion Vol. 4 Drive System Dynamics
   M. Martinez-Sanchez and T. Labuszewski
   MIT Aeroelastic and Structures Lab. Report ASRL TR-184-10

6. Proceedings of NASA Workshop on Wind Turbine Structural Dynamics
   Cleveland Ohio, Nov. 1977
   NASA Publication CONF-771148

7. Proceeding of Second NASA Workshop on Wind Turbine Dynamics
   Cleveland Ohio, March 1981
   (To be published)

Figure 1   Power train details and nomenclature.

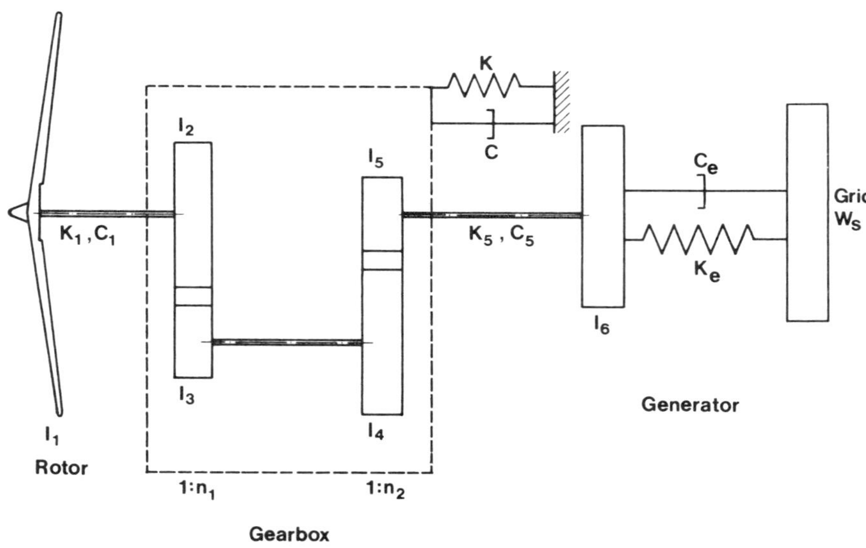

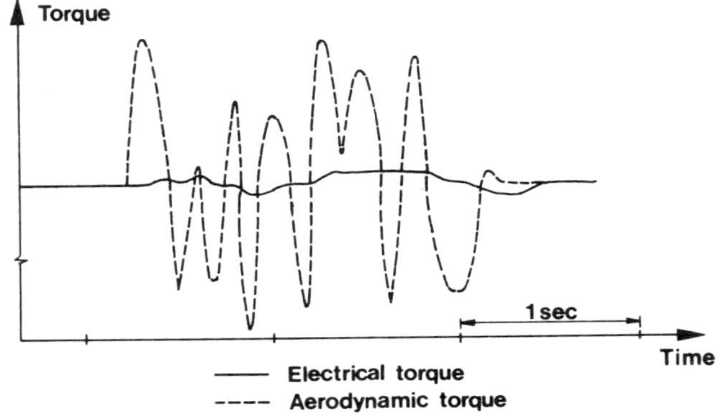

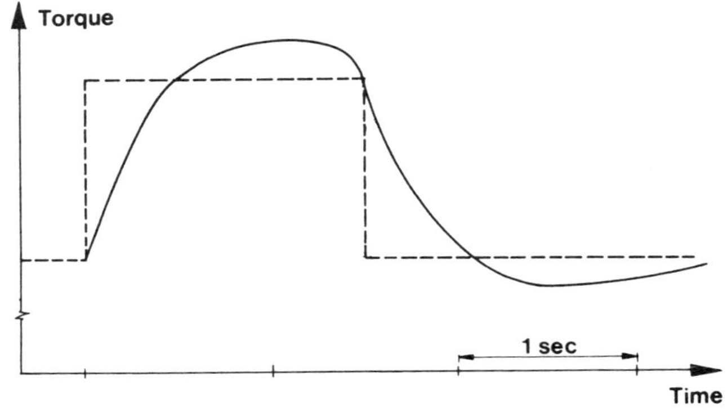

Figure 2   Gust Response

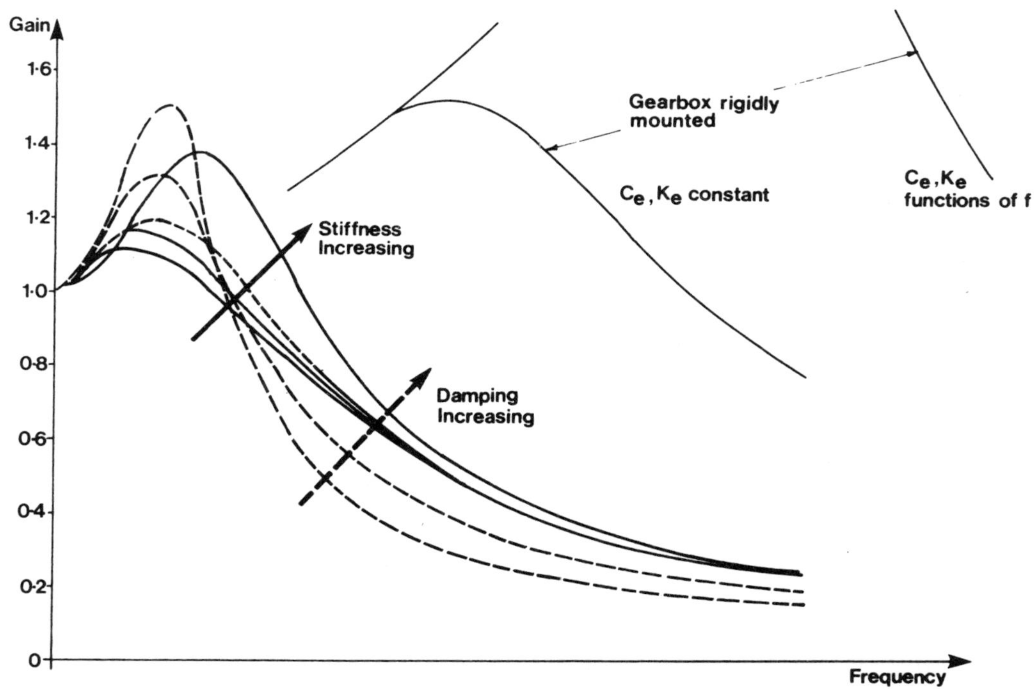

Figure 3 Influence of K and C on gain

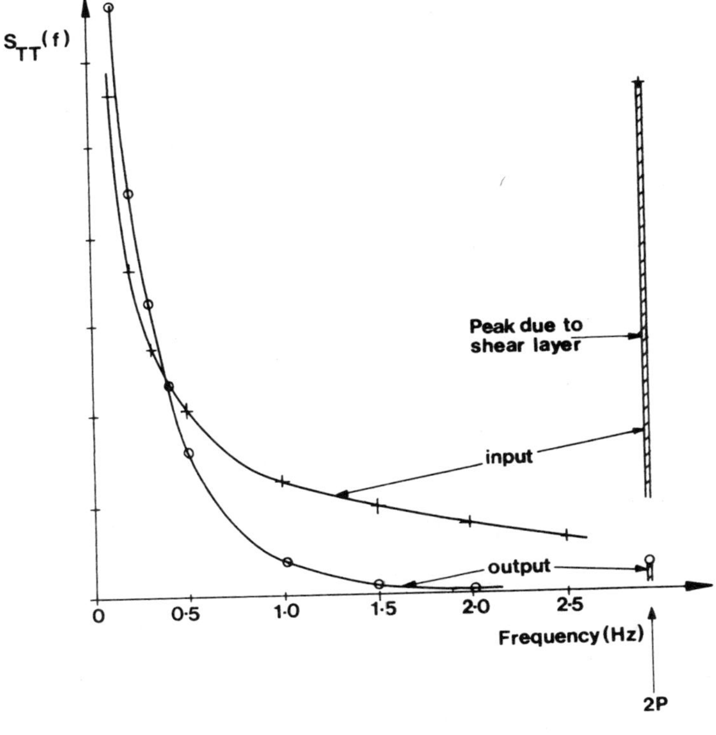

Figure 4
Input and output torque spectra

SWEDYARDS/HAMILTON STANDARD WIND TURBINE SYSTEMS
CHARACTERISTICS OF CONCEPT

Dr A Brännström
Corporate Development
Swedyards Corp
Box 416
401 26 Göteborg SWEDEN

General

The WTS-3 and WTS-4 units as contracted for Sweden and USA, and as ordered for Hawaii, are generally characterized by being dynamically "soft" and by being optimized for energy capture rather than for peak aerodynamic performance. They are designed for fully automatic, unattended operation under remote control.

Turbine

The rotor blades are manufactured from GRP-Energy in a filament winding process. This process is numerically controlled and fully automatic. The winding of the blade is made in two stages. The leading edge D-spar is wound over its mandrel and semi-cured before the trailing edge mandrel is mounted on to the rear end of the D-spar. The winding is then completed over the D-spar and the trailing edge mandrel, giving a continuous and monolithic structure. The mandrels are removed, and the blade is fully cured before final adjustments and fitting of the root end.

The rotor blades are bolted to the teetering hub. The teeter hinge is offset at a certain angle (Delta - 3) to give the blades a cyclic pitch variation, which largely compensates for vertical wind-shear and which - for the same reason - stabilizes the rotor plane in yaw. Consequently, the turbine can be operated downwind free in yaw, yaw motors only used for positioning the turbine correctly downwind in shut-down modes.

Transmission

An important feature in any soft design is the ability of the power-train to absorb torque variations within a certain frequency spectrum. This relieves the control system from some rather unrealistic blade pitch response requirements. In the WTS-3/4 the epicyclic gearbox is mounted with torsional flexibility through the use of springs and dampers. The damping of this movement is very important, as an undamped torsional flexibility can create large phase angle oscillations on the generator side.

Control system

The pitch control system takes account of the Delta-3 teeter hinge offset and the flexible transmission to provide very fast pitch response over the whole span of the rotor. This gives a high degree of optimization of the energy capture over the design wind envelope, beginning from cut-in wind all the way up to rated wind.

The pitch control system is based on microcomputers, and works through two different hydraulic pressure levels. The lower pressure is for all normal operations, while the higher pressure is accumulated to be used only for short "bursts" of control power in gusty winds or in emergencies. Thus, hydraulic pump requirements are kept low.

Tower

The welded steel shell tower is designed as a part of the soft systems dynamics. From the rotor blades, through the hub and transmission and all the way down to the concrete slab under the tower, the aim has been to "uncouple" all main eigenfrequencies, so that the tower foundation - figuratively speaking - does not know what happens to the rotor blade.

The welded tower can be "turned" in its design to fit changing requirements for rotor rpm, gust spectrum etc - within reasonable limits.

Variations in tower height are comparatively cheap in design and production, once the main strength and dynamic properties are established and tested. The welded design lends itself to modular concepts, and to different self-erecting schemes. Tower sections can also be bolted together.

Structure and fatigue life

The basis for a dynamically soft and compliant design must be a very comprehensive systems analysis effort, where load assumptions, dynamics and stresses are iterated several times. Large efforts have been made in this respect for the WTS-3/4 units.

The result is a thouroughly analyzed structure - within the limitations of the knowledge of the load spectrum - with a calculated fatigue life of 30 years.

An interesting detail might be, that the rotor blades are not fatigue limited in their design, as their material thickness is given by stationary buckling loads under some extreme wind conditions. The resulting material thickness gives very low stress levels under highfrequency fatigue loads.

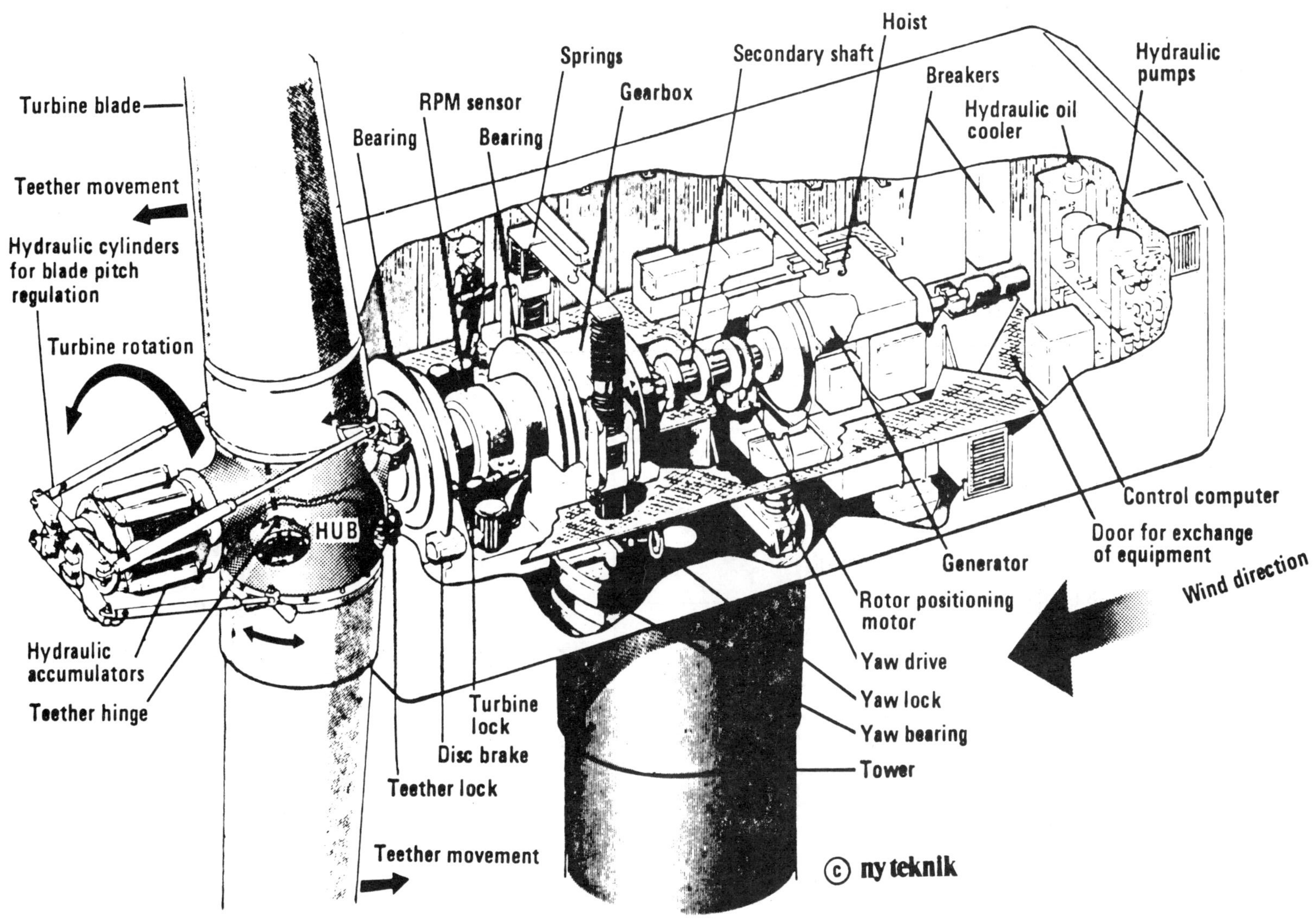

Turbine blade
Teether movement
Hydraulic cylinders for blade pitch regulation
Turbine rotation
Hydraulic accumulators
Teether hinge
HUB
Bearing
RPM sensor
Bearing
Springs
Gearbox
Secondary shaft
Hoist
Breakers
Hydraulic oil cooler
Hydraulic pumps
Control computer
Door for exchange of equipment
Wind direction
Generator
Rotor positioning motor
Yaw drive
Yaw lock
Yaw bearing
Tower
Turbine lock
Disc brake
Teether lock
Teether movement
© ny teknik

WTS-4 FEATURES

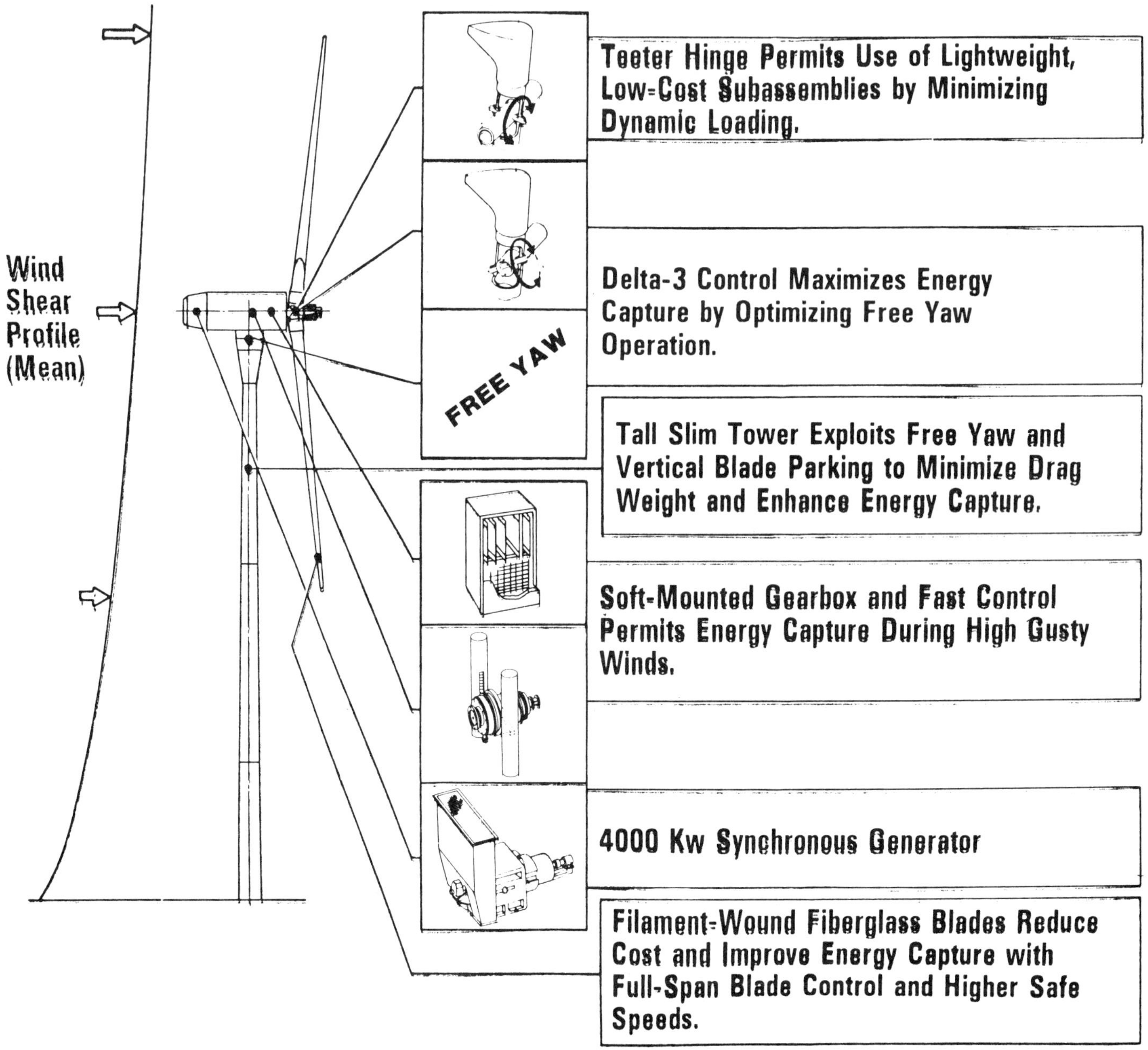

Wind
Shear
Profile
(Mean)

FREE YAW

Teeter Hinge Permits Use of Lightweight, Low-Cost Subassemblies by Minimizing Dynamic Loading.

Delta-3 Control Maximizes Energy Capture by Optimizing Free Yaw Operation.

Tall Slim Tower Exploits Free Yaw and Vertical Blade Parking to Minimize Drag Weight and Enhance Energy Capture.

Soft-Mounted Gearbox and Fast Control Permits Energy Capture During High Gusty Winds.

4000 Kw Synchronous Generator

Filament-Wound Fiberglass Blades Reduce Cost and Improve Energy Capture with Full-Span Blade Control and Higher Safe Speeds.

FILAMENT WINDING PROCESS

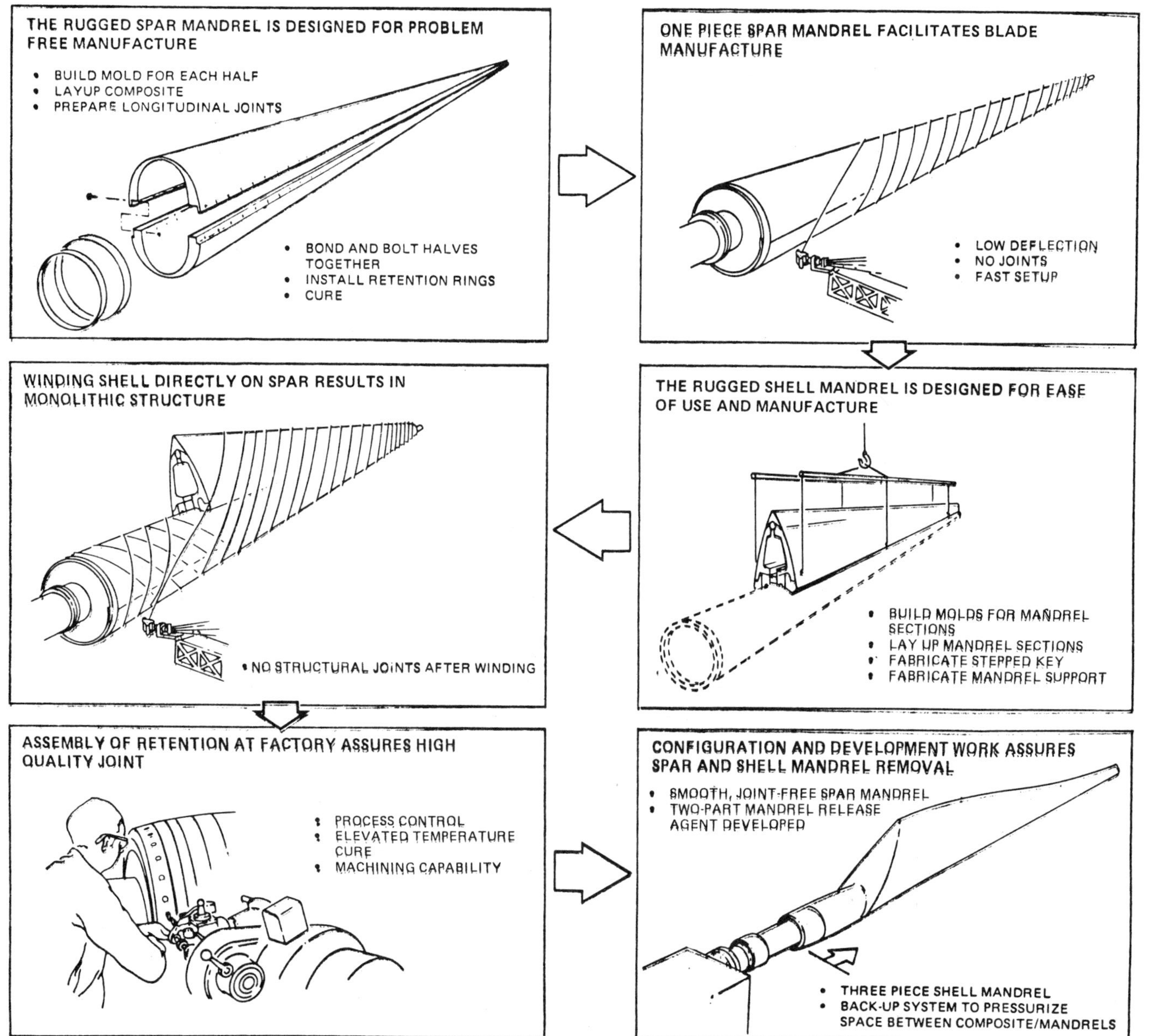

THE 20m DIAMETER WIND TURBINE FOR ORKNEY

J.R.C. Armstrong, Taylor Woodrow Construction Ltd., Southall
G.R. Ketley, British Aerospace Dynamics Group Ltd., Hatfield
B.J. Cooper, GEC Mechanical Engineering Laboratory, Whetstone

ABSTRACT

The 20m Orkney WTG is being built as a test prototype for the 60m, 3MW machine, and as a basis for development of commercial medium size machines. The WTG is a two-bladed, upwind, dynamically stiff design and will be able to run in constant speed or variable speed mode, delivering power into the Orkney grid. Special attention has been paid to the rotational dynamics of the power train, and a soft transmission and feathering blade tips have been designed to achieve high quality electrical power output.

It is intended that the machine will be in operation by Spring 1982. A common data acquisition system will be used to monitor the performance of the 20m and 60m machines.

INTRODUCTION

The design of the 20m wind turbine for Orkney was started in August 1980. The purpose of the machine is to act as a development prototype for the 60m wind turbine, also to be built on Orkney. Experience of designing, building and operating this machine will be of substantial benefit in the development of the 60m machine. In addition it was recognised by the Wind Energy Group (Taylor Woodrow, British Aerospace and GEC) that the 20m machine is of a size ideally suited to many utility applications around the world. WEG is funding the project together with the North of Scotland Hydro Electric Board and the Department of Industry.

The site chosen for the 20m machine is Burgar Hill on mainland Orkney: further details are given by Lindley and Stevenson(1). The machine will feed about 700,000 kWh into the Orkney grid, and is planned to begin operation in Spring 1982. This paper describes the technical features of the installation.

MACHINE DESCRIPTION

The main design drivers for the machine are as follows:

1. Orkney is a very high wind speed site: survival wind gust speeds of 70m/s when stationary and 40m/s when operating are assumed.

2. The design should model the 60m machine, and in view of the importance of the project, should be conservative: accordingly the machine is designed to be dynamically stiff.

3. Although the Orkney supply grid is strong as far as the 20m machine is concerned, there is still a requirement for high quality power. This leads to the adoption of the following three features:

   (i) power trimming rotor
   (ii) soft drive train
   (iii) synchronous generator.

These features have led to the design shown schematically in Figure 1. The main components of the machine are described below.

Rotor

The rotor is two-bladed, and is located upwind of the tower. The outer 20% of each blade can be feathered to 90° to provide power trimming and braking of the rotor. Each blade consists of a fabricated steel spar and an all-enclosing covering of foam and GRP with appropriate twist and taper. This method of construction has been chosen because it appears to be the most cost effective method of providing adequate strength and stiffness for the high wind loading. It also has similarities to that proposed for the 60m design. A rigid hub is employed as it was considered simpler to accommodate alternating stresses by increasing material dimensions than to reduce these by adopting a teetering hinge. Feathering of the tips is accomplished by means of a bell-crank and linkage system which connects the tips to an hydraulic actuator located in the nacelle.

Strain gauges are fitted to the blades and hub of the rotor for monitoring purposes.

Transmission

The rotor shaft is carried by two roller bearings mounted in a fabricated steel housing. The shaft carries at its aft end a two stage in-line gearbox. This is allowed to rotate against the action of springs and dampers. The mechanical compliance introduced in this way allows torque spikes from the rotor caused by gusts to be attenuated before reaching the generator and hence the grid.

An example of how overall power smoothing is achieved is shown in Figure 2. The swinging gearbox is effective in removing high frequency spikes (1a). However, low frequency surges cannot be removed by the transmission alone (1b), and movement of the blade tips is required to spill power and maintain the quality of power output (2). The development and analysis of this system is described in detail by Garrad[(2)].

The high speed gearbox output is connected to the generator by a one-way coupling. This allows the rotor to freewheel below synchronous speed, and prevents power being drawn from the grid during wind lulls. It also eases the synchronisation procedure.

A disc brake is installed on the low speed shaft, and is capable of stopping the rotor from an overspeed condition in high winds with the tips unfeathered.

Nacelle

The power train components are mounted on a stiff steel frame, or pallet, which is in turn mounted on the tower by a standard crane slewing ring. The yaw drive is provided by a highly geared electric motor driving a pinion meshing with the outer gear of the slewing ring. A yaw brake is permanently applied to prevent dynamic instabilities and backlash in the gear teeth.

The nacelle is completed by a floor and housing attached to the pallet. Anemometers, an aircraft warning light and a lightning spike are mounted on the roof of the housing.

Tower

The tower is formed from a cylindrical steel tube mounted below the level of the rotor disc on a conical concrete frustum. Access to the nacelle is provided by an internal ladder leading to an external platform just below the level of the nacelle. A ladder from the nacelle floor provides access from the platform to the nacelle.

The tower foundation consists of an hexagonal reinforced concrete raft which is set without anchorage on the underlying sandstone structure of Burgar Hill.

SYSTEM DESCRIPTION

Controller

The overall WTG system is shown schematically in Figure 3. All functions of the machine are regulated from a control hut located about 90m from the wind turbine. The controller is contained in a single cabinet and carries out two main functions:

1. Supervisory control - start/stop/run.
2. Closed loop regulation of power by tip control.

The supervisory controller scans sensors indicating a number of machine and external conditions - e.g. rotor speed, wind speed - and initiates procedures such as start-up, stop, yaw correction and so on. It is in overall control of the machine, although it can of course be over-ridden manually.

The closed loop controller comes into action when the supervisory controller allows the machine to run and to deliver power to the grid. Its job is to actuate the blade tips via an hydraulic system located in the nacelle, so as to remove power spikes and limit output to the power demanded (maximum 250kW).

Provision is made for the machine to run in either of two modes:

1. Direct connected mode - fixed speed.
2. Power conditioned mode - variable speed.

In direct-connected mode the generator, once it has been driven to synchronous speed by the rotor, is synchronised with the grid and delivers power as the rotor rotates at a constant 88 rpm. The predicted characteristic of power output vs wind speed is shown in Figure 4.

In power-conditioned mode, the rotor is allowed to vary in speed between 44 and 88 rpm so as to follow the characteristic shown in Figure 5. At wind speeds up to 11.5 m/s this represents constant tip speed ratio (= 8); above 11.5 m/s rotor speed is limited to 88 rpm and tip speed ratio falls as wind speed increases. This function is regulated by the closed loop controller, which continuously adjusts the power demanded by the power conditioning unit so as to maintain rotor speed at the appropriate level. As before, the closed loop controller also modulates the blade tips to limit power output to the operating limit demanded (maximum 250 kW).

The supervisory controller is designed to allow automatic operation with a telemetry link to the Board's operating centre. This link provides monitoring of running conditions, warnings and malfunction states (in which case the machine is automatically shut down). The power operating limit can also be adjusted by the Board through the telemetry link.

Testing and Installation

A key feature of the project is the ground test of the complete nacelle and power train assembly at GEC Whetstone. This will be arranged so that a mechanical drive simulating the rotor under various wind conditions will feed power into the turbine shaft, through the transmission and generator and into the grid. Feedback from the actuation of the tips to shaft power input will be simulated by computer.

After the ground tests, the nacelle will be transported to the Orkney site as a single assembly. The hub and blades will be mounted to the shaft and the entire unit lifted and fixed to the tower. Civil works, tower erection and site wiring will have been carried out during the nacelle ground tests.

Data Acquisition

The performance and conditions of the system will be monitored by a data acquisition system which, it is intended, will subsequently be used to monitor the 60m machine. The objectives of the data acquisition are to:

1. Confirm the integrity of the design.
2. Assess the performance of the system.
3. Validate the analytical models used in the 20m and 60m designs.
4. Provide engineering input to the 60m design.

Data will be sampled from four sources:

1. Structural : rotor, nacelle and tower strains and accelerations.
2. Environment : wind, temperature etc.
3. Control system : general conditions - tip angle, rotor speed etc.
4. Electrical : Quality and quantity of power output.

The data will be sampled as follows:

1. Long term : low sample rate over a period of months.
2. Conditions of interest : high sample rate over a period of minutes.
3. Special/commissioning tests : setting up of particular experiments.

All data acquisition will be under the control of a dedicated mini-computer. It is intended that monitoring will be carried out over a period of about one year once the machine is commissioned.

Acknowledgements

The Wind Energy Group acknowledges the support in making this project possible of the North of Scotland Hydro-Electric Board and the Department of Industry.

References

1. D. Lindley and W. Stevenson: "The horizontal axis wind turbine project on Orkney", this Conference.

2. A.D. Garrad: "Wind turbine transmission systems", this Conference.

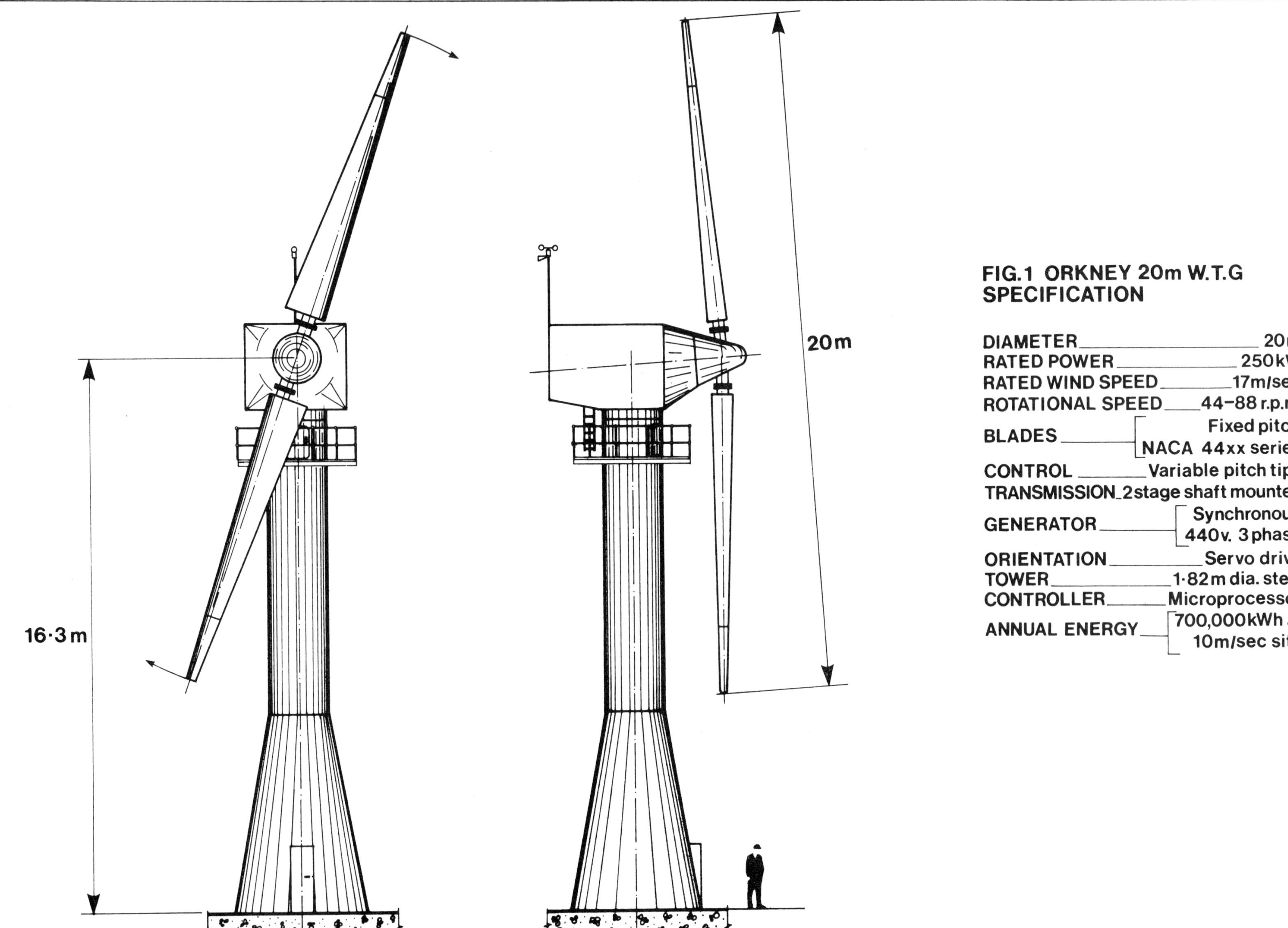

FIG.1 ORKNEY 20m W.T.G SPECIFICATION

Fig.2 20m POWER SMOOTHING

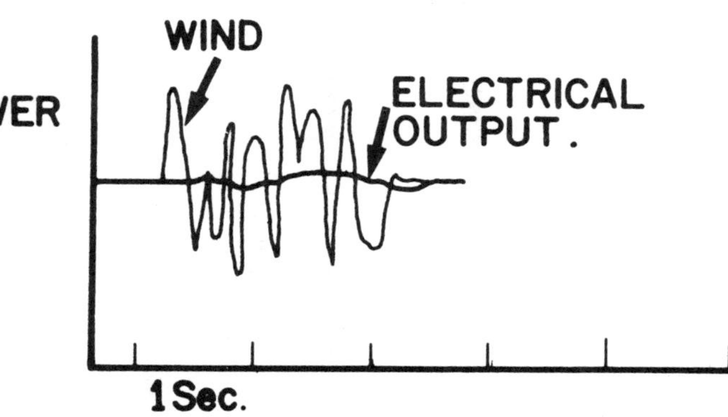

1a. SOFT TRANSMISSION ALONE (NO MOVEABLE TIPS)

EFFECTIVE IN REDUCING HIGH FREQUENCY POWER SPIKES

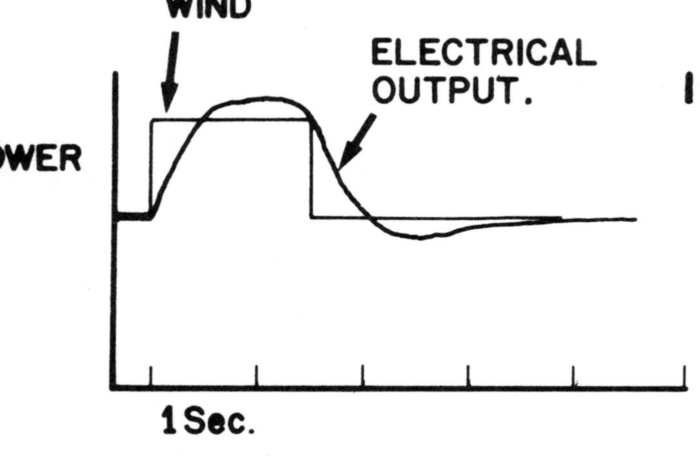

1b. SOFT TRANSMISSION ALONE

INEFFECTIVE IN REDUCING LOW FREQUENCY POWER SURGE.

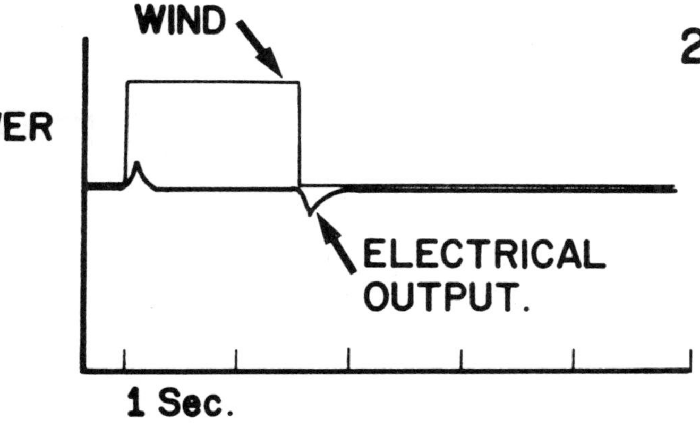

2. SOFT TRANSMISSION PLUS MOVEABLE TIPS

EFFECTIVE IN REDUCING LOW FREQUENCY POWER SURGE.

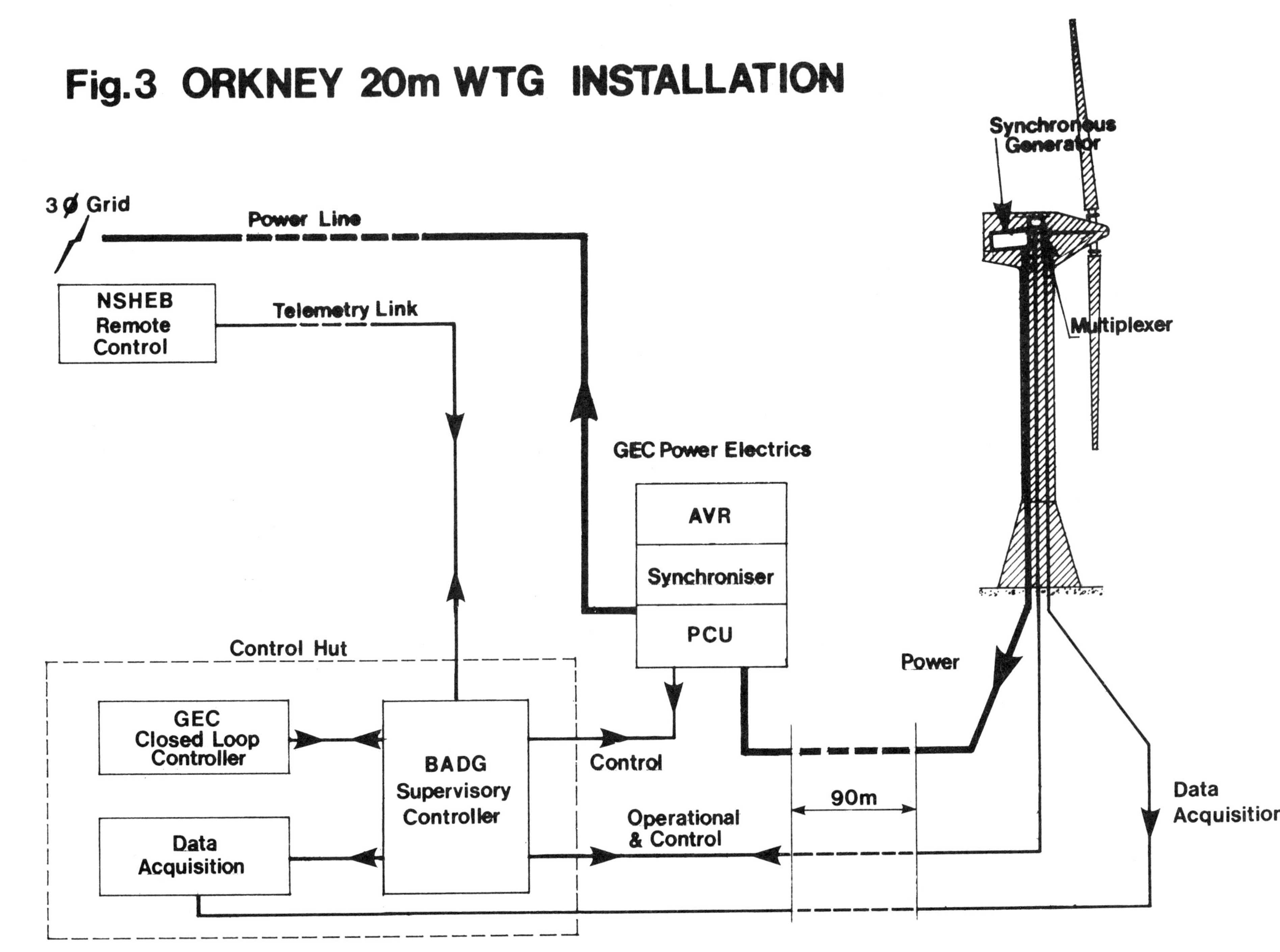

Fig.3 ORKNEY 20m WTG INSTALLATION

Fig.4 20m WTG POWER CHARACTERISTIC

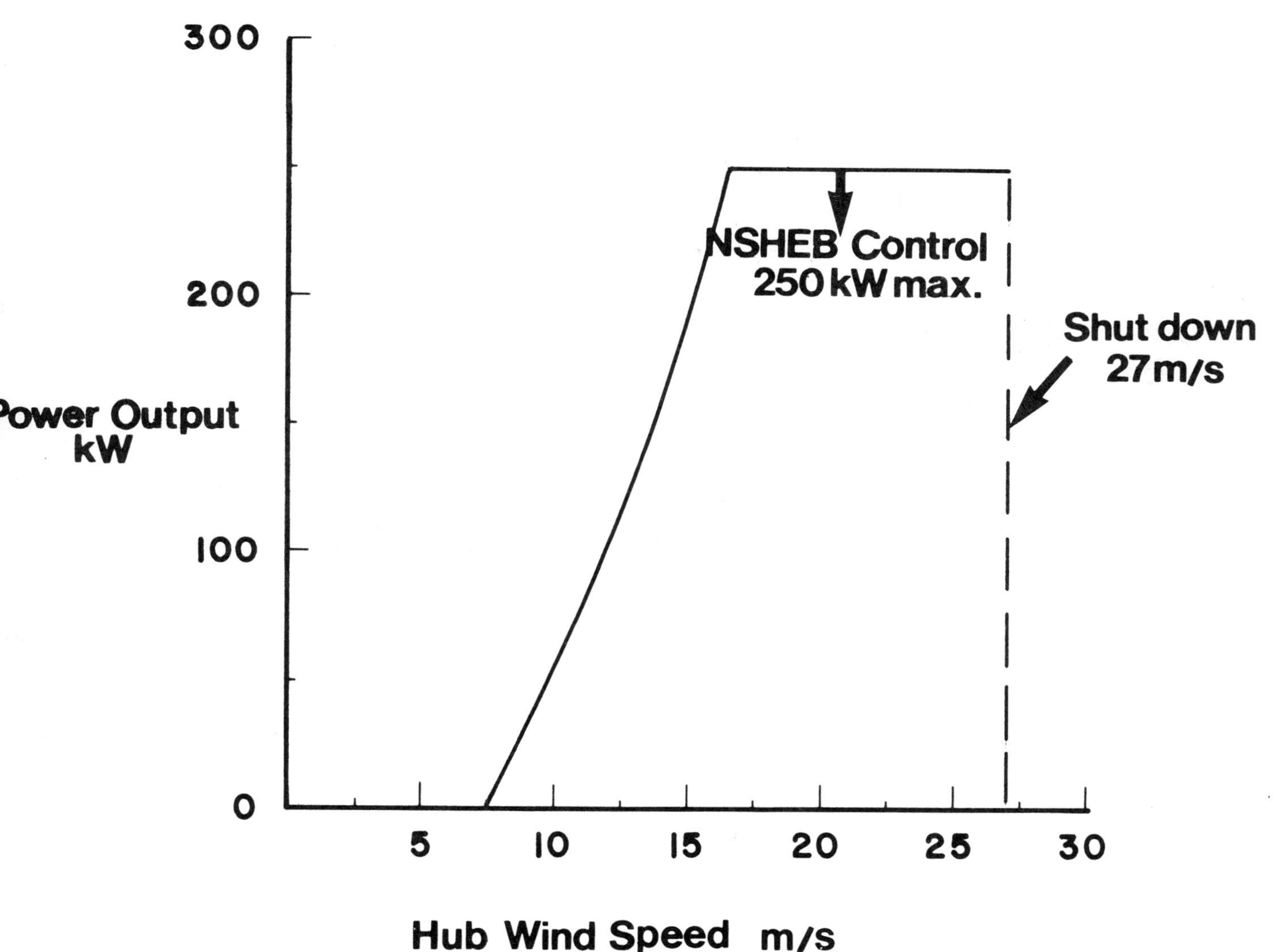

Fig.5 20m WTG SPEED CHARACTERISTICS

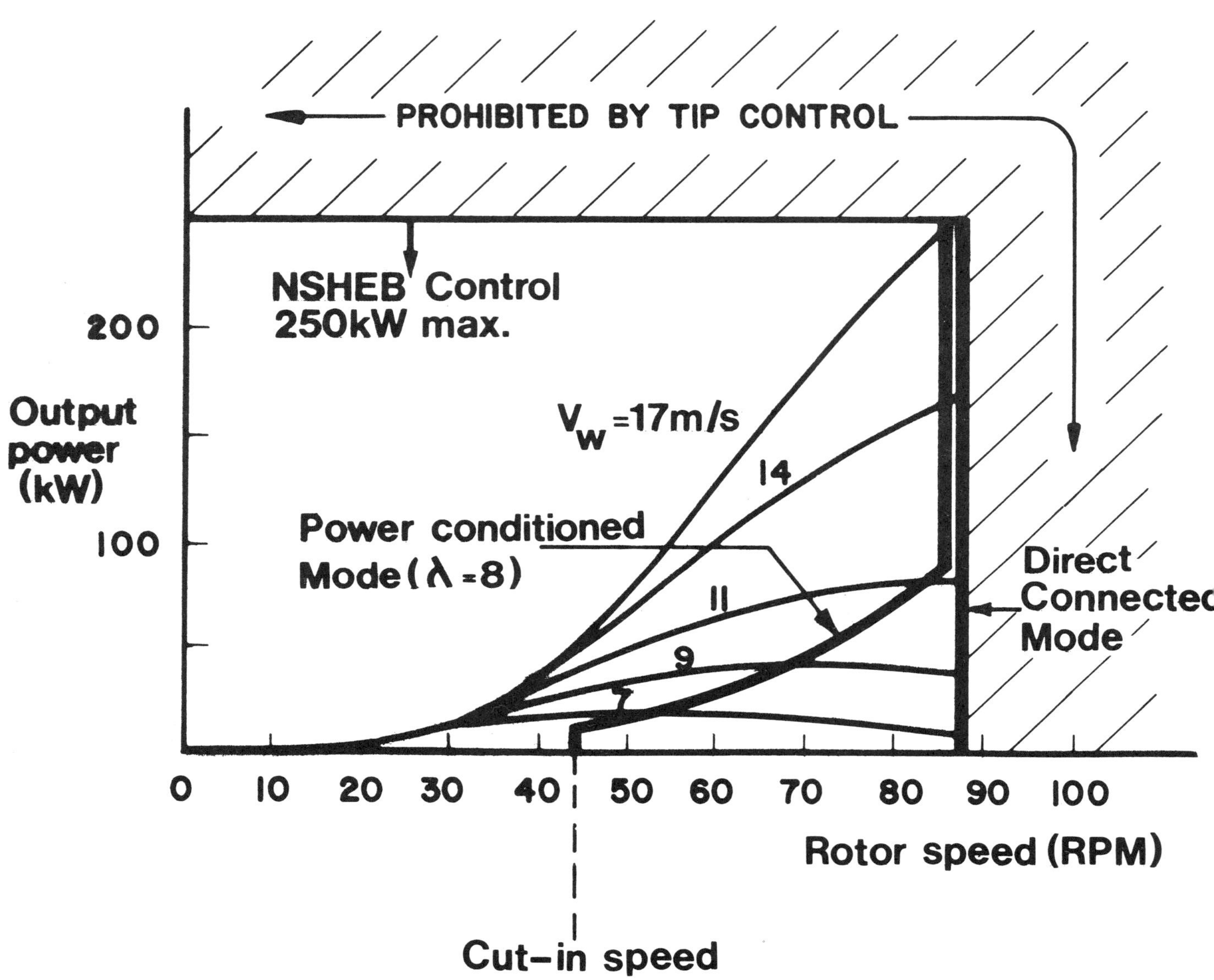

OFFSHORE WIND AND WAVE POWER - A PRELIMINARY ESTIMATE OF THE RESOURCE

A.P. Rockingham, R.H. Taylor, J.F. Walker*
Planning Department
CEGB, London EC1

Abstract

An estimate is presented of the gross wave- and offshore wind power resource around the UK. The wave resource has been calculated from wave data after allowance has been made for device efficiencies and reliability. The gross wind resource has been evaluated for areas further than five kilometres from the shore and for water depths of less than 30 m. Wind conditions, wind turbine characteristics, array efficiencies and reliability factors have been assumed in order to calculate both peak power and annual energy output. The gross resource in both cases will be reduced by a number of factors to give a smaller useable resource. This is estimated in the case of wind power but the paper highlights areas where further investigation will be needed to make a more accurate assessment of this reduction. It is concluded that the gross resource, even on conservative assumptions, is significant in relation to total UK electricity production and is larger for wind power than for wave power.

Introduction

This paper presents the results of a preliminary estimate of the extent of the sea area around the UK which might, in principle, be available for the siting of windmill arrays and wave energy converters. These figures have then been used to calculate the installed capacity and annual energy which would be produced from the areas.

The study for wind power has been performed in two parts.

(i) An estimate has been made of the total surface area within the UK continental shelf boundary of waters less than (a) 25 m and (b) 30 m in depth. A further constraint has been added that no area should be within 5 km of the coastline in order to minimise visual impact from the shore. The total area of these waters has been designated as Class I. An estimate has then been made of the annual energy which might be extracted using present designs of wind turbine arranged in large clusters.

It should be made clear that both of these constraints have been defined arbitrarily and warrant further investigation. Building wind turbines in deeper water would clearly increase the available resource but, at present the sensitivity of cost to increased water depth is not well defined. Evidence from recent US studies indicates that building in deeper water may not in fact increase costs significantly. There is little understanding of the environmental effect of wind turbines and it is possible that the 5 kilometre constraint is unnecessarily severe.

* Present address: Research Division, The Marchwood Engineering Laboratories.

(ii) The total sea area designated as Class I has been reduced to attempt to take some account of sea-bed conditions and such factors as shipping lanes and fishing areas. It is to be emphasised that the resulting area (designated, Class II) should only be regarded as providing a very tentative estimate of the area which might be available for development. Should windmill clusters require to be built offshore, detailed siting studies would need to be carried out and these notional areas may be considerably reduced. The annual energy available from Class II areas has then been calculated.

Results - Windpower

1. Estimates of Areas

Table 1 shows the total offshore area with depths less than 25 m and 30 m respectively and further than 5 km from the shore in each of the above classes. The locations of these areas are shown in fig. 1. The areas around Flamborough Head, Happisburgh/Maplin and Morecambe Bay have been reduced by a factor of the order of 50% to obtain Class II figures. This reflects the density of fishing and/or shipping in these areas. Because of high shipping density, the area along the south coast has been neglected. The area of the Dogger Bank has been excluded because of the unsuitable sea-bed conditions in this area.

Including the 29 dispersed sites less than 150 km^2 in area, a total gross potential (Class I) of 33,000 km^2 is obtained. With the allowance mentioned above, this area is reduced to 14,000 km^2 (Class II). Comparable figures for areas up to 25 m depth are 17,000 km^2 (Class I) and 7,500 km^2 (Class II).

2. Available Wind Power

The following assumptions with regard to annual mean wind speeds, wind turbine characteristics and array configuration and efficiencies were made:

(i) Wind Speed

Examination of isovent maps for UK offshore areas suggests an average annual mean wind speed over the areas considered of 6.5 m/s at a height of 10 m. Assuming a height scaling exponent of 0.1, this gives an annual wind speed of 8 m/s at a hub height of 80 m above mean sea level.

(ii) Wind turbine characteristics

A machine with 80 m diameter blades rated at twice the mean wind speed has been assumed. The peak output of this machine would be 2.5 MW, and the load factor 31%.

(iii) Array Configuration

An array efficiency of 60% given by Milborrow (1979) for an "infinite" array with 10-D spacing has been assumed. This may be a conservative figure since not all areas will be so large as to be regarded as infinite in the context of Milborrow's results.

(iv) Reliability

Studies (Walker, 1981) indicate that the reliability of the mechanical and electrical components of the wind power clusters would combine to give overall reliability of between 85% and 95%. A central value of 90% has been assumed.

Using these data, a peak power density of 5 MW/km^2 and an energy density of 7.2 GWh/km^2 is obtained. The total power and annual energy available from Class I and Class II areas for each water depth is summarised below.

	Class I		Class II	
	25m	30m	25m	30m
Peak Power (GW)	84.6	166.	38.3	69.1
Average Power (GW)	14.1	27.7	6.4	11.5
Energy p.a. (TWh p.a.)	124.	243.	56.1	101.

If the mean wind speed at hub height is increased from 8 m/s to 10 m/s the power and energy outputs in all cases are approximately double those given above. This sort of sensitivity analysis highlights the need for careful consideration of site windspeeds.

In the Offshore Study (Simpson and Lindley, 1980) windspeeds at a height of 80 m were estimated to be 8.9 m/s, 9.2 m/s, and 8.6 m/s (all with an error bound of ± 1.2 m/s) for Morecambe Bay, Camarthen Bay and the Wash Approaches respectively. The estimates in this study are thus conservative.

Results - Wavepower

In the case of wavepower, the devices are assumed to be installed along the lines shown in fig. 1.

The table below shows the length of line assumed for each of the three locations, the average incident power in kW/m including a reduction to take into account directionality, the capacity of typical devices installed along each line, the annual average power output and the number of TWh generated per annum. In order to obtain the peak power levels involved, a device peak output of three times the mean output has been assumed (Glendenning, 1978). Device productivity (capture efficiency x power chain efficiency) is taken to be 33%. A reliability factor of 90% was assumed.

The offshore areas chosen and the figure for average incident power in the waves are taken from Winter (1980). It is possible that these figures underestimate the convertible power available in the sea. Recent sea trials on a particular type of wave energy converter (Count 1980) have suggested that the directionality factors may be somewhat higher than those assumed by Winter, perhaps by 20-40%. By contrast, however, the reliability factor of 90% may be an overestimate. It is unlikely, therefore, that these uncertainties will cause any dramatic change in the power and energy estimates and so the final conclusions would be unaltered.

	Length of line	Average incident power (including directionality)	Installed Capacity	Average Power	Energy p.a.
	km	kW/m	GW	GW	TWh
Shetlands-Hebrides	820	29	23.7	7.1	62.1
SW Approaches	190	25	3.4	1.4	12.3
North East	450	6	1.5	.8	7.0
Totals	1460		31.1	9.3	81.5

The figures taken represent an estimate of gross potential which might be compared with the wind resource designated as Class I. No reductions have been made to take account of sea-bed conditions and such factors as shipping lanes and fishing areas.

Discussion

The resource has been described using both energy and power figures with some trepidation. Both figures can be used incorrectly. A maximum power figure can provide useful information if systems integration problems are to be studied but it does not relate in any simple manner to the peak load that can be satisfied. For example, a high maximum power figure, and a low annual energy figure indicate that back-up costs associated with the use of wind- and wave-power could be large. If only an average power figure were to be quoted, while the resource size would be more meaningful for comparison with conventional plant, information on the variability of the resource would be lost.

This study is, by necessity, extremely superficial in a number of respects. To improve it, further investigation is required of:-

(a) the relationship between water depth and the cost and feasibility of support structures for aerogenerators;

(b) the constraints on siting which could exist due to ship movements, fishing and spawning grounds and defence installations. This is relevant for both wind and wave energy;

(c) wind and wave characteristics at possible sites;

(d) interaction between machines which might reduce cluster efficiencies;

(e) the characteristics of arrays of wave energy converters with particular reference to directionality;

(f) economic and technical optimisation of wind and wave generators.

The present study has not addressed itself to economic issues. Further studies might be aimed at obtaining estimates of the available resource within defined cost ranges.

Conclusions

On conservative assumptions with regard to wind speeds and array efficiencies the gross wind power resource in waters less than 30 m in depth is likely to be about 165 GW of installed capacity representing an annual energy of about 240 TWh p.a. (comparable with current UK electricity generation). If very rough allowance is made for reduction in siting area for fishing, shipping and sea-bed unsuitability these figures are reduced to about 70 GW and 100 TWh.

The gross wavepower potential is estimated to be about 31 GW and 81 TWh p.a.

Although various constraints may significantly reduce figures for both wind- and wavepower, it appears that the gross resource (should it be required to be developed) is much higher for wind- than for wavepower, although both are likley to be very significant.

Acknowledgement

This paper is published by permission of the CEGB.

References

1. Milborrow, D.J. The Performance of Arrays of Wind Turbines, Journal of Industrial Aerodynamics, 5, 1980, 403-430.

2. Winter, A.J.B. The UK Wave Energy Resource. Nature, Vol. 287, October 1980, pp 826-828.

3. Simpson, P.B., Lindley, D. Offshore Siting of Wind Turbine Generators in UK Waters, British Wind Energy Association Workshop, April 1980.

4. Glendenning, I., Generation and Transmission, Proceedings Wave Energy Conference, London, November 1978.

5. Count. B., Private Communication, 1980.

6. Walker, J. Estimation of the Availability of an Offshore Wind Power Installation, British Wind Energy Association Workshop, April 1981.

TABLE 1

Class I and Class II Areas for Wind Power

	Area (km^2)			
	Class I		Class II	
	25m	30m	25m	30m
East Coast				
1. Dogger Bank	1,000	6,000	-	-
2. Flamborough Head (Outer Wash)	5,400	11,000	2,700	5,500
3. Happisburgh) Maplin)	3,300	5,000	1,500	2,000
South Coast				
4. Isle of Wight) Dungeness)	1,000	1,100	-	-
West Coast				
5. Carmarthen Bay	300	400	300	400
6. Cardigan Bay	1,000	1,500	1,000	1,500
7. Morecambe Bay (Solway, Liverpool, Isle of Man, Lancaster Sands)	3,100	4,800	1,500	2,200
8. Luce Bay	1,200	1,300	200	300
Dispersed Sites (each less than 150 km^2)	700	2,200	500	2,000
Total	17,000	33,300	7,700	13,900

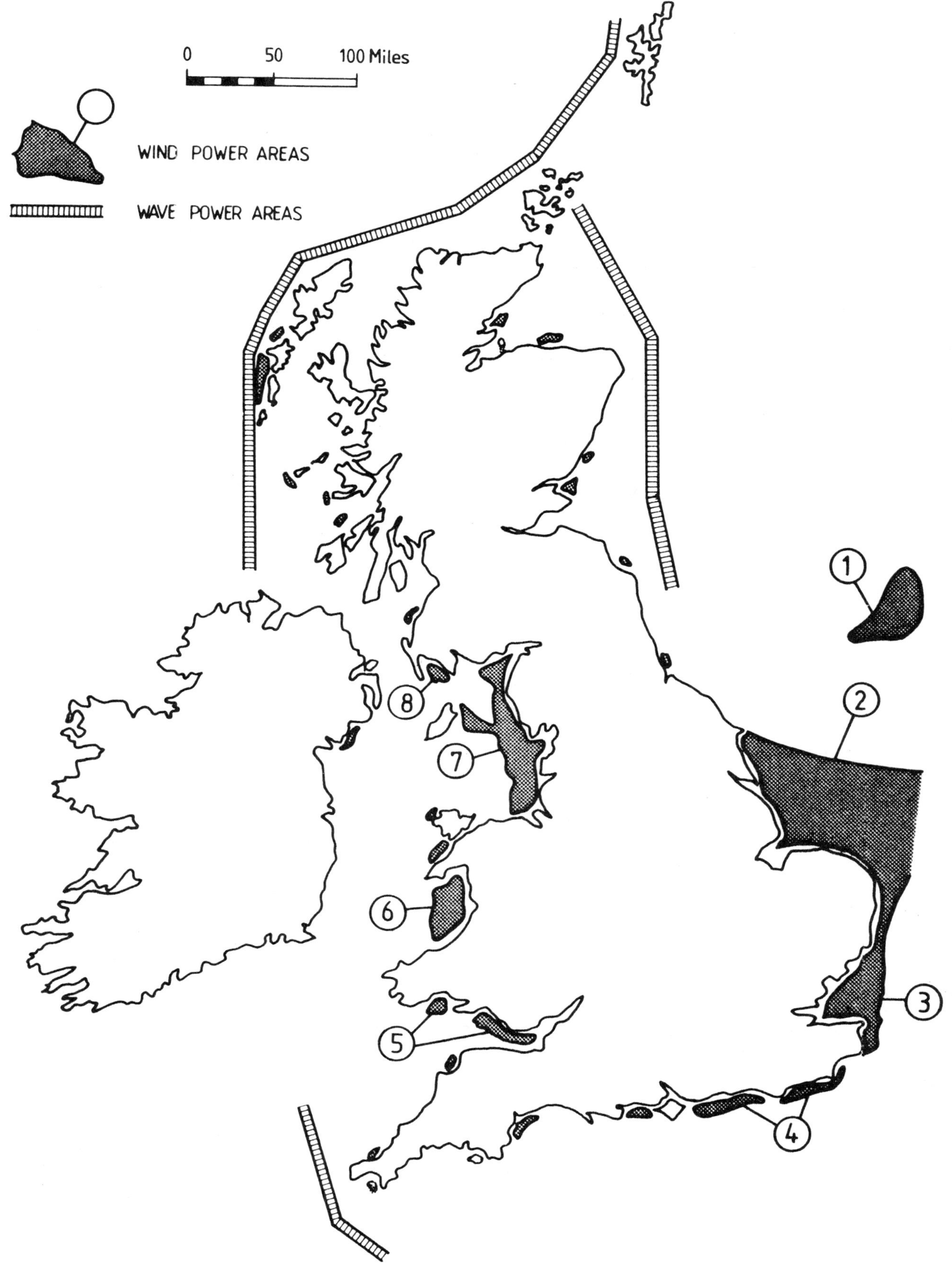

Figure 1 OFFSHORE WIND AND WAVEPOWER AREAS OF INTEREST

ESTIMATION OF THE AVAILABILITY OF AN OFFSHORE WIND POWER INSTALLATION

J.F. Walker
Headquarters Planning Department*
Central Electricity Generating Board
London

Abstract

A method is described for estimating the annual availability of an offshore wind power installation, containing 196 wind turbines of maximum output 3.7 MW, given mean times to failure and repair of the main components, including the wind turbine generators, their transformers, switchgear, cabling and transmission links to the shore. The effects are considered of alternative transmission arrangements.

It is shown that, using a range of failure and repair times, the system would be available to transmit power if required for about 85-95% of the time. The limitations of the method and of the data are considered.

Introduction

This paper describes a method of quantitatively assessing the availability of an offshore wind power system and discusses some of the implications of the results. The work was part of the contribution by the Central Electricity Generating Board to the Taywood Consortium study carried out in 1979 for the UK Department of Energy (Simpson et al., 1).

The basic arrangement considered is a 14 x 14 array of wind turbines, each of maximum output 3.7 MW, with two 400 kV submarine connections to the shore as shown in Figure 1. The effects are considered of alternative 132 kV transmission arrangements as shown in Figure 2.

The sensitivity of overall availability to repair and failure times of transmission equipment is taken into account by consideration of two cases, the first using a basic set of repair times derived from sources given below and the second using repair times which are half those of the first.

Two repair situations have been considered. The first assumes that all underwater equipment and transformers have a mean time to repair of 3 months, and switchgear 1 week. All repair times are longer than those given by the statistics for land based equipment. In the second case, all electrical repair times were assumed to be half those of the first case.

These data and the assumptions on which they are based are discussed more fully below. Derived failure rates and repair times are listed in Tables 1 and 2.

*Present address: Central Electricity Generating Board, Marchwood Engineering Laboratories, Marchwood, Southampton, SO4 4ZB.

Maintenance and Reliability Assumptions

With regard to the individual wind turbines, it is assumed that some types of failure will be repairable in the winter. Others will not be repaired until the summer, when a campaign of repairs will be required. This will restore each failed machine to as new condition and will also serve as a major 5 year overhaul if the machine is due for one.

During the summer, which is assumed to be a period of 6 months, routine electrical and mechanical overhaul will be required on all wind turbines which have neither had a major repair nor are due for a 5 year overhaul.

Two types of summer operating failures are considered, requiring major and minor repairs. Summer wind speeds and hence operating stresses will generally be lower than in the winter. The summer failure rate is therefore assumed to be half that of winter.

Data on winter major failure rate, major overhaul, repair and annual overhaul times and number of repair crews are as derived in the Taywood Study. Other data were obtained by intelligent guesswork. Values of mean time between failure (mtbf) and mean time to repair (mttr) are listed in Table 1.

Failure data on 11 kV, 132 kV and 400 kV equipment were obtained from Electricity Supply Industry sources and are summarised in Table 4.

It has been assumed that there is no difference between summer and winter repair times, based on CEGB experience of the existing dc Cross Channel link. The year has therefore been taken as a whole, rather than being split into two as when considering the wind turbine generators.

Method

Fundamental assumptions are that failures are random and independent and failure rates constant. The system is assumed to be fully developed, that is, the probability of being in any particular state does not change with time. The method is generally probabilistic and is based on techniques such as those described by Billinton (2) for estimating the availability of power systems.

Because of constraints on space, the full derivations are not given of equations for the component and group availabilities, although the basic equations are stated and in some cases the derivation is obvious.

Basic Definitions

Consider any period of time T in which an item of equipment is unavailable for a number of periods totalling time t. Availability of the item:

$$A = (T - t)/T \tag{1}$$

If failures and repairs are random and independent and failure rates λ and repair rates μ are constant then:

$$A = \mu/(\lambda + \mu) \tag{2}$$

The reciprocals of λ and μ are the mean time between failure (mtbf) and mean time to repair (mttr) respectively.

If the total number of items of requipment is N and their availability is A, then the number available C is given by:

$$C = NA \qquad (3)$$

Wind Turbine Winter Availability

If $\lambda_{2.1}$, $\mu_{2.1}$ are the winter minor failure and repair rates respectively, then the availability $A_{2.1}$ of wind turbines not including the effect of major failures is as follows:

$$A_{2.1} = \mu_{2.1}/(\lambda_{2.1} + \mu_{2.1}) \qquad (4)$$

If $\lambda_{2.2}$ is the winter major failure rate, T is the length of the winter period, the number of available machines n in a population initially containing N is:

$$n = Ne^{-\lambda_{2.2}T} \qquad (5)$$

The availability of the wind turbines not including the effect of minor failures is given by:

$$A_{2.2} = 1 - A_{2.1}[e^{-\lambda_{2.2}T}/\lambda_{2.2} + T - 1/\lambda_{2.2}]/T \qquad (6)$$

The overall winter availability of wind turbines by elementary probability theory is:

$$A_2 = A_{2.1} \cdot A_{2.2} \qquad (7)$$

Wind Turbine Summer Availability

Summer availability A_3 from equation 1 is

$$A_3 = 1 - (T_{3.1} + T_{3.2} + T_{3.3} + T_{3.4})/T_3 \qquad (8)$$

The individual times unavailable are given by equations 9-13:

$$T_{3.1} = p_{3.1}j(1 + j/q)/2 \qquad (9)$$

where $T_{3.1}$ is the total time unavailable during the summer for repair of winter failures. j unrepaired failures are waiting at the start of the summer for repair by q repair crews, i.e. q failures can be repaired concurrently with a mean time to repair of $P_{3.1}$.

$$T_{3.2} = p_{3.2}(N - j)i/k \qquad (10)$$

where $T_{3.2}$ is the total time unavailable for major overhauls. We assume that the total economic life of the wind turbines is k years, the mean time between overhauls is i and that the j machines which fail during the winter are restored to as new condition during the repair period $p_{3.1}$

$$T_{3.3} = p_{3.3}(N - j)(1 - j/k) \quad (11)$$

where $T_{3.3}$ is the total time for annual mechanical and electrical overhauls, which are required for all machines which have not had a major repair or overhaul.

$$T_{3.4} = (T_3 - T_{3.1} - T_{3.2} - T_{3.3})/(1 + 1/\lambda_{3.4}p_{3.4}) \quad (12)$$

where $T_{3.4}$ is the total time unavailable due to summer failures, T_3 is the product of the total number of machines and the length of the summer period, $\lambda_{3.4}$ = summer failure rate and $p_{3.4}$ = mean summer time to repair.

$$T_3 = NT \quad (13)$$

$\lambda_{3.4}$ and $p_{3.4}$ are given by equations 14 and 15:

$$\lambda_{3.4} = \lambda_{3.4.1} + \lambda_{3.4.2} \quad (14)$$

$$p_{3.4} = (\lambda_{3.4.1}\mu_{3.4.2} + \lambda_{3.4.2}\mu_{3.4.1})/(\lambda_{3.4.1} + \lambda_{3.4.2})\mu_{3.4.1}\mu_{3.4.2} \quad (15)$$

where λ and μ are failure and repair rates; 3.4.1 and 3.4.2 relate to major and minor failures.

Wind Turbine Generator Annual Availability

Since the winter and summer periods are of equal length, annual availability A_1 is the arithmetic mean of the winter and summer availabilities.

$$\therefore \quad A_1 = (A_2 + A_3)/2 \quad (16)$$

Wind Turbine Group Availability

The average number of machines available during the year in a group (see, for instance, Billinton (2)), is given by the following:

$$C_7 = A_6 \sum_{r=1}^{s} (A_{1.r}A_{4.r}A_{5.r}) \quad (17)$$

The number of machines in each group is s; $A_{1.r}$, $A_{4.r}$ and $A_{5.r}$ are respectively the availabilities of the rth machine, its switch and the 11 kV connection to the 11 kV-132 kV transformer. A_6 is the availability of the transformer. Individual availabilities A_4, A_5 and A_6 are determined using the appropriate failure and repair rates and equation 1. The availability of a machine group is as follows:

$$A_7 = C_7/s \quad (18)$$

Transmission Availability - 400 kV Links to Shore

The availability of the individual transmission components are determined using the appropriate failure and repair rates and equation 1. The availability of 132 kV cables and equipment associated with each machine group is as follows:

$$A_{12} = A_8 A_9 A_{10} A_{11} \tag{19}$$

where A_8, A_9, A_{10} and A_{11} are the availabilities of the cable, 132 kV switches, 132-400 kV auto transformer and 400 kV isolator respectively.

The availability of one 400 kV substation cable is as follows:

$$A_{15} = A_{13} A_{14} \tag{20}$$

where A_{13} and A_{14} are the cable and switch availabilities respectively. All substation cables are assumed to be identical.

The availability of one 400 kV link to the shore is given by:

$$A_{18} = A_{16} A_{17} \tag{21}$$

where A_{16} and A_{17} are the cable and switch availabilities respectively. All 400 kV links are assumed to be identical.

Overall System Availability - 400 kV Links to Shore

The following assumptions are made:

An outage of a 132 kV cable causes an outage of all the machine groups connected to it. Both the 400 kV substation cable and the 400 kV link to shore are capable of carrying the entire output of the system, i.e. there is full redundancy.

By elementary probability theory, overall system availability A_{22} is therefore:

$$A_{22} = A_7 A_{12}(1 - (1 - A_{15})^2)(1 - (1 - A_{18})^2) \tag{22}$$

The capacity available of the complete system is given by equation 5.

Overall System Availability - 132 kV Links to Shore

Making the following assumptions:

(i) an outage of a 132 kV cable between the transformer and the interconnecting cable causes an outage of all the machine groups connected to it;

(ii) if a 132 kV cable to shore is unavailable, one alternative route only is possible, which is capable of carrying its normal load and that of the failed link.

Then, by elementary probability theory

$$A_{22} = A_7 A_{12}(1 - (1 - A_{15})^2) \tag{23}$$

In this case there is no 132-400 kV transformer and switch, so A_{10} and A_{11} in equation 21 (which gives A_{12}) are both unity.

Results

The results of the availability calculations are presented in detail in Tables 1, 2 and 3.

Discussion

The wind turbine failure rates were not based on failure analysis of the failure modes of the various subsystems, which was not possible in the time available. They were obtained by intelligent guesswork based on engineering experience. Repair and maintenance times were derived after consideration of the time required for each class of maintenance activity. The validity of these assumptions is therefore uncertain, although they can be taken as a reasonable starting point.

Failure rates for electrical equipment were derived directly from data for land based equipment, no changes being made for the transition from land to offshore. There is some justification for this assumption; Dawson et al. (3) assumed failure rates of 2/100 km year for unburied submarine cables after consideration of data from a number of sources. The failure rates quoted in Table 4 are all higher than this value and therefore may be conservative.

The mean time to repair of 3 months which has been assumed for underwater cables is also thought to be conservative. By comparison, Dawson et al. (3) assume the mean time to repair of hvdc cables to be 1000 hours, which is considerably less than 3 months. Given the availability of suitable repair vessels, it should be possible to achieve mttr's considerably less than 3 months even taking into account delays due to adverse weather.

The assumed mean times to repair for transformers were 3 months, which compare with about 8 days and 35 days mttr respectively quoted in Table 4 for 132 kV and 400 kV land transformers.

The mttr of 7 days for all switchgear compares with 5-10 hours mttr for 11 kV switches, about 6 days for 132 kV equipment and 2 days for 400 kV equipment.

The method of estimating availability is based on probability theory and on techniques such as those described by Billinton (2) which have been shown to give results of acceptable accuracy when applied to conventional power systems. Although the method deals only with the steady state, it could be adapted to cover the years leading up to the time when the first five year overhauls were due.

The assumptions that failures are random and independent and that failure and repair rates are constant may give rise to error, which should be given consideration in any future work.

The results must be considered in the light of the reservations expressed above concerning reliability and maintainability data and of the theoretical basis of the method. For the long mttr case annual availabilities are 86% for the 400 kV and 90% for the 132 kV arrangement, whilst halving the mttr increases the availabilities to 90% and 93% respectively. Exactly the same effect would have been produced if the failure rates had been halved. This shows that both failure and repair rates must be determined with adequate and equal accuracy.

Availability is sensitive to the availability of the transmission link to shore, an improvement of 4% in the long mttr and 2% in the short mttr case being obtainable by the use of two 132 kV links instead of two 400 kV links.

The results for the system with 132 kV links to shore are conservative because it is assumed that there is only one alternative route available in the event of failure of a link. Reference to Figure 2 shows that there are actually three, although the availability of additional links depends on the power being generated at any particular time. At periods of high power the additional links would not be capable of taking the additional load. The additional availability which would result from what is in effect partial triplication would be about 1-2% for the high mttr case. However, the same level of improvement can be obtained by halving failure rates or repair times. The reasons for the improved availability of the 132 kV arrangement excluding the effects of triplication are that there are no 132-400 kV transformers and 132 kV cables are more reliable than 400 kV cables, as shown in Table 4.

The results of the availability studies have demonstrated the sensitivity of the overall availability to failure and repair rates and to the system configuration. Availabilities as high as 95% might possibly be achieved with high quality engineering design, operation and maintenance. A credible range of availabilities that could be considered to determine the sensitivity of overall economics to availability is 85%-95%.

Acknowledgements

This paper is published by permission of the Central Electricity Generating Board.

References

1 Simpson, P.B. Lindley, D., Hardy, W.E. An Assessment of Offshore Siting of Wind Turbine Generators in the United Kingdom. Proceedings of Third International Conference on Future Energy Concepts, IEE London, January 1981. Conference Publication 192. 264-272.

2 Billinton, R. Power System Reliability Evaluation. Gordon and Breach (New York) 1970. 299 pp.

3 Dawson, J.M., Din, S., Mytton, M.G., Shore, N.L., Stansfield, H.B. System Reliability Studies for Wave Energy Generation. Proceedings of Second International Conference on Future Energy Concepts, IEE London. January 1979. Conference Publication 171. 171, 170-176.

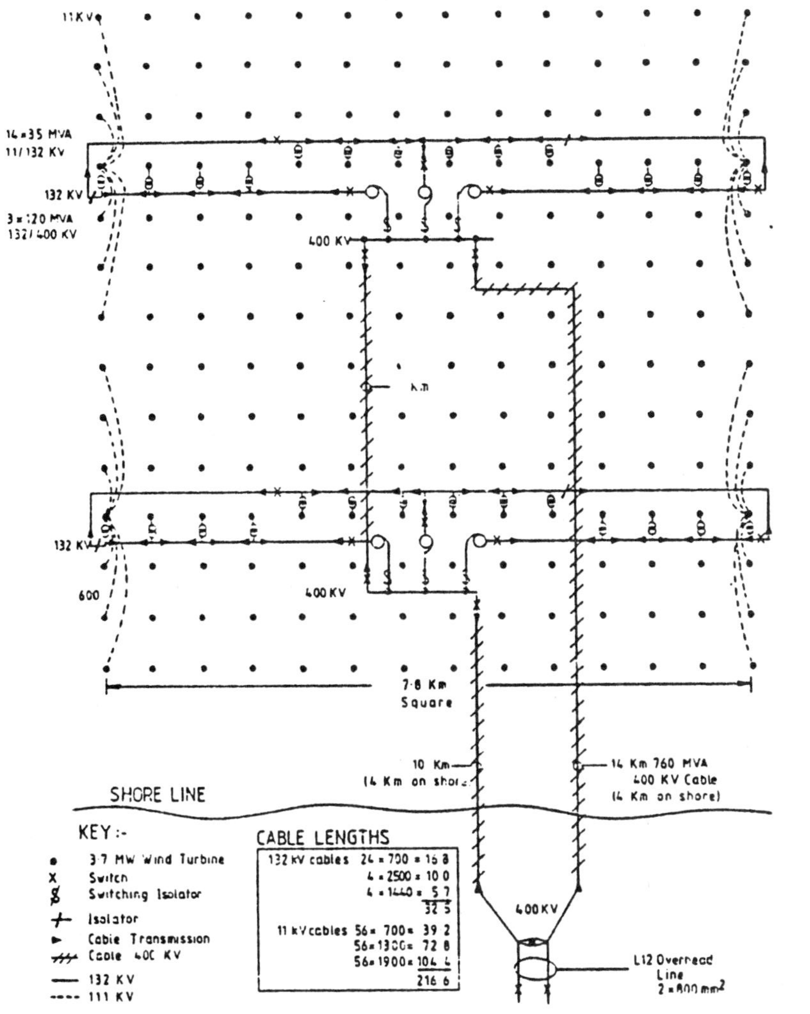

WIND POWER ELECTRICAL TRANSMISSION 400 KV OFF SHORE CONNECTIONS
14×14 Array 3·7 MW Turbines = 725 MW

FiG.1

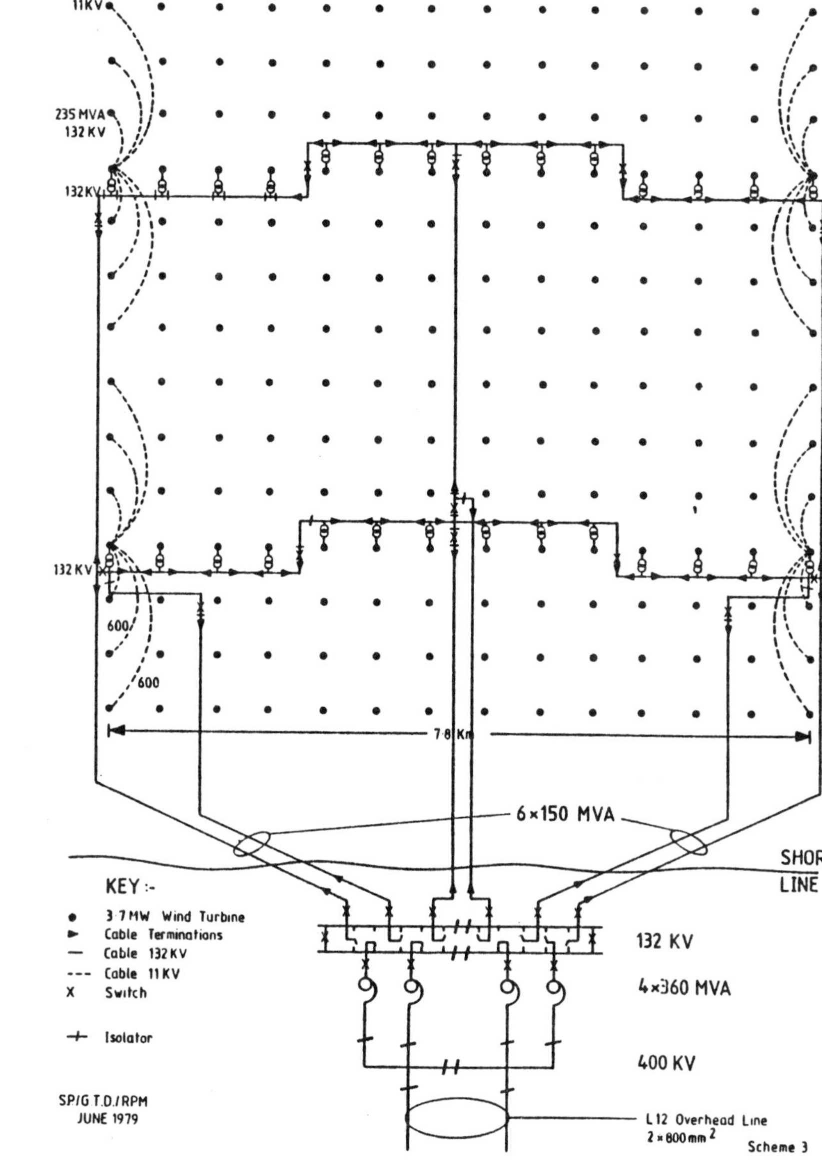

WIND POWER ELECTRICAL TRANSMISSION 132 KV OFF SHORE CONNECTIONS
14 × 14 Array 3·7 MW Turbine = 725 MW

FIG 2

TABLE 1

WIND TURBINE GENERATOR WINTER, SUMMER AND ANNUAL AVAILABILITIES

	Failure rate λ no/yr	Mttr days 1/μ	Time available or unavailable years	Availability
Winter				
Minor (repairable) failures	$\lambda_{2.1}$ 4	$1/\mu_{2.1}$ 1	$T_{2.1}$ 1.015	$A_{2.1}$ 0.99
Major (unrepairable in winter) failures	$\lambda_{2.2}$ 0.167*	-	$T_{2.2}$ 3.979	$A_{2.2}$ 0.96
Major and minor failures			4.994	
Total machine time			T_2 98	
Overall winter availability				A_2 0.95
Summer				
Repair of winter failures		$P_{3.1}$ 10	$T_{3.1}$ 1.622	
Major 5 year overhaul		$P_{3.2}$ 10	$T_{3.2}$ 0.986	
Annual mechanical and electrical overhaul		$P_{3.3}$ 3.5	$T_{3.3}$ 1.381	
Repair of summer failures	$\lambda_{3.4}$ 0.083	$P_{3.4}$ 10	$T_{3.4}$ 1.213	
Total summer hours unavailable			5.202	
Total machine time			T_3 98	
Overall summer availability				A_3 0.95
Annual availability				A_1 0.95

* The failure rate $\lambda_{2.2}$ = 0.167/yr corresponds to a 92% survival probability over 6 months

TABLE 2

COMPONENT AND SYSTEM AVAILABILITIES - 400 kV LINKS TO SHORE

	No. of items in group	Cable length km	Failure rate λ no/yr	Repair rate low: Repair rate μ no/yr	Repair rate low: Availability	Repair rate high: Repair rate μ no/yr	Repair rate high: Availability
Wind turbine generator	7		See table 1		A_1 0.95	table 1	A_1 0.95
Switch	7		0.004	52	A_4 1.00	104	A_4 1.00
11-132 kV transformer	1		0.036	4	A_6 0.99	8	A_6 1.00
Individual cables	2	1.9	0.128	4	A_5 0.97	8	A_5 0.98
	2	1.3	0.088	4	A_5 0.98	8	A_5 0.99
	2	0.7	0.047	4	A_5 0.99	8	A_5 0.99
	1	0.2	0.014	4	A_5 0.99	8	A_5 1.00
Wind turbine generator group					A_7 0.92		A_7 0.94
132 kV cable	1	3.25	0.081	4	A_8 0.98	8	A_8 0.99
132 kV switch	2		0.022	52	A_9 1.00	104	A_9 1.00
132-400 kV transformer	1		0.091	4	A_{10} 0.98	8	A_{10} 0.99
400 kV isolator	1		0.205	52	A_{11} 0.99	104	A_{11} 1.00
Overall availability of cable & equipment					A_{12} 0.95		A_{12} 0.98
400 kV sub cable	1	0.25	0.012	4	A_{13} 1.00	8	A_{13} 1.00
400 kV isolators	2		0.410	52	A_{14} 0.99	104	A_{14} 0.99
Overall availability of sub stations					A_{15} 0.99		A_{15} 0.99
Availability of one or both sub station					A_{20} 1.00		A_{20} 1.00
400 kV cable to shore	1	14	0.662	4	A_{16} 0.86	8	A_{16} 0.92
400 kV switches	2		0.616	52	A_{17} 0.99	104	A_{17} 0.99
Overall availability of cable & equipment					A_{18} 0.85		A_{18} 0.92
Availability of one or both cables					A_{21} 0.98		A_{21} 0.99
Availability of complete system					A_{22} 0.86		A_{22} 0.91

TABLE 3

COMPONENT AND SYSTEM AVAILABILITIES - 132 kV LINKS TO SHORE

	No. of items in group	Cable length km	Failure rate λ no/yr	Repair rate low		Repair rate high	
				Repair rate μ no/yr	Availability	Repair rate μ no/yr	Availability
Wind turbine generator group (as for table 2)					A_7 0.92		A_7 0.94
132 kV sub cable and switch*					0.98		0.99
132 kV cable to shore	1	9.76	0.244	4	A_{13} 0.94	8	A_{13} 0.97
Switches	2		0.008	52	A_{14} 1.00	104	A_{14} 1.00
Overall availability of cable & equipment					A_{15} 0.94		A_{15} 0.97
Availability of one or both cables					A_{20} 1.00		A_{20} 1.00
Availability of complete system					A_{22} 0.90		A_{22} 0.93

* Equipment is the same as that in Table 2 but excluding 132-400 kV transformer and switches

TABLE 4

SUMMARY OF BASIC ELECTRICAL FAILURE RATES IN FAILURES/YEAR

1	11 kV switch	0.35/100 switches
2	11 kV cables	6.73/100 km
3	11-132 kV transformers	3.6/100 transformers
4	132 kV cable	2.5/100 km
5	switches	1.1/100 switches
6	132-400 kV auto transformer	9.1/100 transformers
7	400 kV cable	4.73/100 km
8	400 kV switches & isolators	20.52/100 items

BWEA 3rd W E C Cranfield 1981

LARGE OFFSHORE WIND TURBINES: SYSTEM DESIGN AND ECONOMICS

J C Dixon
Open University, UK

Abstract

There have been four principal studies of the prospects for large offshore wind turbines to generate electrical energy in large quantities - in UK, US, Sweden and Holland. The UK has a particularly good offshore wind resource. The advantages and disadvantages of offshore and onshore machines are compared and it is concluded that offshore can be economically competitive and has some environmental advantages. The baseline designs from the UK, US and Swedish studies are compared. The three designs for UK conditions (according to each of the three studies) are compared and evaluated economically according to appropriate UK accounting procedures. The energy cost estimates are found to be about 1.8, 2.8 and 4.2 p/kWh with a mean value a little under 3 p/kWh. The dependence of the costs of support platforms on depth are described and it is concluded that the extra costs for deeper water could be offset by quite a modest increase in mean windspeed.

The costs of grid connection are analysed. There are considerable discrepancies between the estimates in the three studies. However although longer distances from the shore do increase costs signifiantly, this is not overwhelming, and again would be compensated by a fairly small increase in mean windspeed. It is finally concluded that 2 p/kWh (1980 prices) is a realistic target at which to aim, for machines to be built in large quantities in ten years or so. Target component costs are suggested, totalling on investment cost of 500 £/kW, to meet that energy cost target.

Introduction

This paper highlights and extends some of the results of a study of offshore wind energy (ref.1)conducted for the recent BWEA Position Paper (2) concentrating particularly on economics implications.

The UK has an excellent offshore wind resource although our detailed knowledge of offshore wind behaviour is poor, particularly in terms of the accuracy needed for good engineering design and economic assessment (3,4). Even when considering only the resource over shallow waters (<30 m) with a notional reduction for shipping and other uses, the potential yield is about 12 GW average (100 TWh/y), almost half the current CEGB average output (5). For practical purposes the resource is 'infinite', and implementation would actually be limited by the CEGB's ability to integrate the variable output into the system economically. Without considerable advances in storage methods or reoptimisation of plant mix this limit is considered to be about 20% of installed capacity (12 GW in 1980, Ref2).

Offshore to land-based comparison

There are a number of major advantages and disadvantages of offshore turbine location compared with the land-based equivalent.

Advantages:
(1) Increased mean windspeed and hence increased energy production from a given rotor size.
(2) Possibility of larger unit size because of an increase in component transportation size limits.
(3) Considerably reduced land use and aesthetic impact.
(4) Greater diversity of location may allow higher capacity credit.

Disadvantages:
(1) Increased size of supporting structures to allow for water depth and resistance to wave action.
(2) Worse survival wind conditions.
(3) Additional electricity transmission cost to shore.
(4) The need for marinisation (resistance to marine corrosion conditions).
(5) More difficult maintenance.

Apart from the land-use and aesthetic considerations, the comparison is primarily an economic one. Table 1 shows typical estimates of the economic effect on the cost of energy (p/kWh) arising from the main factors. The increased cost of the support structure can be considerable and is sensitive to marine conditions, particularly water depth, survival waves and seabed geology. The lower value of +20% only applies to very favourable sites such as the shallow and sheltered Baltic Sea. For harsher North Sea conditions the effect may be nearer to the top of the range. Allowing for both advantages and disadvantages, the conclusion is that offshore wind generated electricity is highly competitive with flat land sites, although probably not competitive with the limited number of favourable 'hill-top' sites, predominanty in Scotland. If a large total power installation is sought, then the reduced land use and reduced aesthetic impact of offshore installation may in any case prove decisive.

Baseline Studies

Offshore studies of substance have been published for UK (3,4,6,7) USA (8,9) and Sweden (10,11). Table 2 shows the comparative data for the baseline designs of horizontal axis machine. The economic figures refer to prices at January 1980. The US figures include 28.5% inflation from the originally published data, and were exchanged at 2.27 S/£ (12). The Swedish figures include 16.2% inflation and were exchanged at 9.45 Kr/£ (12). The UK figures include 10% inflation (July 1979 to January 1980) from those originally published.

The delivered energy cost estimates (table 2 item 25) seem to show quite good agreement between UK and USA, with a remarkably low value for the Swedish system. These comparative economic figures are however misleading for several reasons. There are substantial discrepancies between the three studies in that they do not all take account of array efficiency, transmission efficiency and availability. The environmental conditions also vary considerably. The US baseline design case is 160 km offshore in 150 m deep water and subject to 26 m waves and 1.8 m/s current. In contrast, the Swedish design in the sheltered Baltic is only subject to 10 m waves, and is established in much shallower water (20 m) and is only 20 km from the coast, although it is disadvantaged by a lower mean windspeed. These discrepancies are significant cost drivers.

There is a further, non-technical, and perhaps less obvious difference between the studies which is actually the most important in terms of predicted energy cost: the accounting basis is different from country to country.

UK Conditions comparison

The appropriate way to make a comparison between the studies is to standardise on realistic UK environmental conditions and economic criteria. Table 3 shows the result, in which the three systems have been placed 20 km offshore in 15 m water depth with a 9.5 m/s mean hub-height windspeed, an array efficiency of 84%, transmission efficiency 90%, availability 90% and 20 year life.

The US design utilises a downwind rotor, and for 15 m depth has a steel gravity base caisson with steel lattice tower, in contrast to a tension leg platform for 150 m depth. The capital cost saving is about 87 £/kW (217 £/kW reducing to 130 £/kW for the platform, see section 5 ,figure 2). Reduction of the offshore transmission distance (160 km to 20 km) saves 73 £/kW for undersea cable (section 6 figure 5) and a further estimated 10 £/kW laying cost. The total saving is then 170 £/kW.

It is more difficult to determine an adjustment to the Swedish estimate, because the principal change is an increase in the severity of wave loads. The total capital cost has been increased by 30% to allow for this. The Swedish design has an upwind rotor with cylindrical steel upper tower and concrete lower tower and base.

A further minor discrepancy between the costings should be mentioned. Only the UK study included the cost of transmission from the shore station to an appropriate point on the grid. This is typically 5 to 15 £/kW, which is only about 1% of the total. Within the accuracy of the figures being dealt with here this is probably negligible, but so that the costs quoted can be considered to adequately represent the complete installation, 10 £/kW has been added to both the US and Swedish estimates giving the final figures of 549 £/kW and 456 £/kW respectively.

Incorporation of array efficiency, transmission efficiency, availability and standardisation of the hub-height mean windspeed all affect the system output and the final system plant factor. The plant factors shown in Table 3 are based on the total system power installed, rather than the maximum shoreline power delivery as was used in the original US report (8). The high plant factor for the Swedish design results from a low rated speed of 15 m/s, giving Vr/Vm = 1.6. This suggests that some economic improvement would result from reoptimisation, to allow for the higher mean widspeed figure here (9.5 m/s) than for the original design (8.7 m/s). For example, the cost sensitivity analysis of the UK study (7) suggests that the difference could be a reduction in energy cost by just over 5% from that shown, to 1.7 p/kWh.

The typical system plant factor of 0.25 based on installed power corresponds to about 0.33 based on the peak power seen at the shoreline because of array and transmission efficiencies. These are quite low figures by power station standards.

The final estimated energy cost depends upon the discount cash flow (dcf) rate chosen, and also on assumptions regarding the installation period of the system. A full calculation is rather detailed, so the method used here is to apply a fixed charge rate (fcr) of 7.45% which was chosen as it gives the same final energy cost as the full calculation on 5% dcf rate for the UK system. The actual dcf rate applicable in any particular case is an 'administrative' decision that would be arrived at by discussion between the CEGB and the Government, but 5% appears to be an appropriate and likely value for assessment of renewable energy sources. The capital charge portion of the total energy cost estimate is given by c = Cr/ph where c (p/kWh) is the capital charge C (£/kW) is the capital cost, r is the charge rate (%), p is the system plant factor and h is 8760 hours/year. There has been no attempt here to correct for differing material and labour rates prevailing in the different countries.

The final energy cost estimates of Table 3 range from about 1.8 through 2.7 to 5.6 p/kWh. Notwithstanding some differences in maintenance costs (1), the delivered energy cost is primarily dependent on the capital cost, but the reason for the wide range of capital costs is not clear. The UK platform and tower design is entirely concrete. The US design for this depth has a four cornered steel tube lattice tower on a sheet steel gravity caisson, which might have a low initial cost but could suffer from maintenance problems. The Swedish design uses a cylindrical steel upper tower with concrete lower tower (in the corrosion prone splash zone) and concrete caisson, and has been planned for a high production rate of a hundred or more per year for several years, which could give substantial economies of large scale production.

If the UK design could be built in existing shipyards, as planned for the Swedish design, there would be a capital saving of 63 M£, i.e. 0.31 p/kWh leaving 5.28 p/kWh. The most significant remaining difference is probably in the rotor size. Increasing the rotor diameter from 80 m to 100 m saves 1.0 p/kWh giving a cost estimate of 4.18 p/kWh (following the sensitivity analysis of (7)). The particular diameters investigated by the various national programmes are in fact somewhat arbitrary; for example the US 106 m is really the round figure of 350 feet. The UK sensitivity analysis suggests that the energy cost is roughly inversely proportional to rotor diameter. Using this relationship to adjust to a standard 100 m diameter gives the estimate of Table 4. The mean value of 2.95 p/kWh arguably represents the near state-of-the-art cost, although the hardware remains to be demonstrated. The upper and lower estimates are now about +-40% of the mean. This range of disagreement cannot be considered surprising in view of the complexity of the design subject and range of possible assumptions regarding production methods, quantities and efficiencies.

The UK sensitivity analysis indicated that an extension to 40 year life could reduce final energy cost by 18%. The mean estimate then becomes 2.42 p/kWh, and the lowest (Swedish) one only 1.50 p/kWh. Figure 1 shows the mean cost estimate dependence upon diameter (inverse proportion, following the UK analysis), extrapolating somewhat incautiously to 140 m diameter (although larger diameters are known to be under serious consideration, e.g., West German Growian). One reason that very large unit size looks particularly attractive for offshore applications is that wave forces become less significant.

Support platform costs

In the UK and Swedish studies, the support platform is physically integrated with the tower, and the costs are not easily separated out. However in the US study they are clearly indentifiable. A number of platform types were costed, each over a range of depths. The most economic one for shallow waters (<40 m) was the steel gravity base. In deeper waters the tension leg platform was best. The spar-buoy type showed promise but suffered from a high mooring maintenance cost. Figure 2 shows the platform cost found as a function of depth. The range studied was up to 500 m but the range of interest for UK domestic application is limited to rather less than 200 m. Two important features to be noted from figure 2 are the high peak in platform cost at 40 m depth, and the declining cost beyond that depth (the tension leg platform cost is minimum at 300 m).

From figure 2 it is clear that it should not be assumed too hastily that shallow water sites are best. For example the cost at greater than 150 m depth is actually less than that at 30 m. The tendency for mean windspeeds to increase offshore (typically 0.5 m/s per 10 km for UK) must further enhance interest in more distant offshore sites. As an example, figure 3 shows the variation in mean windspeed and the delivered energy cost as a function of offshore distance for the coast of New Jersey (derived from Ref 8).

Much of the UK offshore falls in the 30-100 m depth range and hence in the cost peak of the curve of figure 2, and therefore it is of great interest to examine alternative structural methods for this depth range. In fact the US report leaves some interesting alternatives unexamined over this range. The concrete gravity-base platform was only considered for depths over 60 m (200 ft), costing typically three-quarters of the steel gravity base type for a given depth. The reason for limiting consideration to depths over 60 m is not given, but an extrapolation of the cost result to shallower waters (figure 4), as a first approximation, would be in agreement with the UK and Swedish studies, which favoured a concrete gravity design.

The US study considered the concrete pile platform for very shallow waters (up to 15 m, 50ft). It was more expensive than the steel gravity-base, but relatively insensitive to depth. Again no reason was given for the depth range limitation, but a tentative extrapolation of its cost looks interesting (figure 4).

The US Study was the only one to examine floating platforms (including the TLP) in detail. However it did not attempt to develop a true optimum design, but rather performed a 'local' optimisation of platform designs taken from oil industry experience. This suggests that further design improvements and cost reductions should be possible. In summary, the design of support platforms appears to be a particularly open question with considerable scope for development.

Grid connection costs

Figure 5 shows typical grid connection costs as a function of the offshore distance of the array, based primarily on figures from the US study (8). The costs fall into three main parts:

(1) Collection cost from individual machines to the array substation
(2) Undersea cable from the array to the shore receiving station
(3) Land connection into the grid (mostly overhead cables)
Also included is a point costing calculated from the Swedish studies (10). For close offshore the UK study estimate was typically 180 £/kW (3).

DC transmission uses cheaper cable than AC but requires rectifier and inverter equipment. The breakeven point is about 74 km for the typical case shown, so a UK system would probably use AC transmission. The gradient of the cost line shows that the cable cost is about 0.85 £/MW m for AC and 0.32 £/MW m for DC. The Swedish studies show a cost dependence on transmission distance which corresponds to a similar AC cable cost. In contrast, figures in the UK study indicate a minimum cost of 2.5 £/MW m for 132 kV cable and 2.0 £/MW m for 400 kV cable The reason for these discrepancies is not clear, but may be related to quantity ordered.

Increasing the offshore distance from 20 km to 100 km results in a capital cost increase of 53 £/kW, for the same depth. Even for a well developed system costing 500 £/kW, this increase is only about 10%, and would be economically justified by a mean windspeed increase of 0.5 m/s or more.

Target costs

In view of the various offshore studies available in the literature, for machines to be built in the later 1980s a target delivered energy cost of 2 p/kWh seems an ambitious but appropriate target. Table 5 shows a likely breakdown of that cost, which would require a capital investment of about 500 £/kW. Table 6 shows a capital cost target list to meet that total, and to be realised in quantity series production. The Swedish cost estimates are below this figure, even with the increase for more severe UK conditions, and if extended to 40 year life give a final energy cost of only 1.5 p/kWh.

Conclusions

The mean of three estimates for the delivered energy cost from 100 m diameter horizontal axis wind turbines in near UK offshore conditions is 2.95 p/kWh. With further development, conventional two-blade horizontal axis machines show promise of reducing costs down to or below 2.0 p/kWh. There are a number of major innovative ideas, not yet adequately assessed, which may improve upon the conventional basic machine (vertical axis, asymmetric rotors, biplane blades etc).

Offshore wind energy has higher energy density than flat land sites which approximately offsets the increased capital costs. Moving further offshore or into deeper water adds further cost, but at a rate which might be justified by reaching higher mean windspeeds.

The economics of offshore wind energy are likely to benefit particularly from improved knowledge regarding the offshore wind, wave prediction, wave force estimation, gigacyle fatigue, protective coatings and operating experience; and from improved designs in respect of flexible towers, platform design for all depths, site preparation and installation, long lifetime, low maintenance, larger rotors, vertical axis rotors and large scale production.

It seems possible that by the end of the century the delivered energy cost from offshore wind will be below that from coal-fired stations.

References

1. Dixon J C, Offshore wind turbines: a review of progress, Faculty of Technology, Open University, UK 1980.
2. BWEA, Wind Energy for the Eighties, Peter Peregrinus, 1981.
3. Simpson P B and Lindley D, Offshore siting of wind turbine generators in UK waters BWEA Second Wind Energy Conference, Cranfield UK Multiscience 1980.

4. Assesment of offshore siting of wind turbine generators, Report for the UK Department of Energy, Taywood Engineering Ltd December 1979 (Unpublished)
5. CEGB Annual Report and Accounts 1979-80
6. Lindley D, Simpson P B, Hassan U and Milborrow D, 'Assessment of offshore siting of wind turbine generators' BHRA Third International Symposium on Wind Energy Systems, Denmark August 1980.
7. Simpson P B and Lindley D 'An assesment of offshore siting of wind turbine generators in the United Kingdom' Third Future Energy Concepts Conference, IEE 1981.
8. Kilar L A, Design study and economic assessment of Multi-unit offshore wind energy conversion system application', Final report Westinghouse Electric Corporation for US DOE, ERDA contract no E(49-18)-2330 June 1979.
9. Kilar L A and Chowaniec C R 'A technical and economic assessment of offshore wind energy conversion systems' Summer Power Meeting, Vancouver BC July 1979.
10. Hardell R and Ljungstrom O 'Offshore based wind turbine systems for Sweden - a system concept study', BHRA second international symposium on wind energy systems, October 1978.
11. 'Wind power at sea' National Swedish Board for Energy Source Development June 1979 (unpublished).
12. United Nations Monthly Bulletin of Statistics

TABLE 1 Offshore effect on cost of energy

1. Increased Vm	0 to -60%
2. Unit size	0 to -40%
3. Tower and platform	+20% to +100%
4. Shore transmission	+10%
5. Marinisation	+4%
6. Maintenance	+10%

TABLE 2 Baseline cost studies (January 1980 prices)

			UK	US	Sweden
1.	Mean windspeed (hub)	m/s	9.5	10.6	8.7
2.	Max wind gust (hub)	m/s	57	98	51
3.	Hub height (asl)	m	80	80	85
4.	Water depth	m	15	150	20
5.	Max wave (50 year)	m	20	26	10
6.	Max current	m/s	1.2	1.8	-
7.	Offshore distance	km	20	160	20
8.	Rotor diameter	m	80	106	102
9.	Rated power	MW	3.73	10	7
10.	Rotor loading	W/m2	742	1133	857
11.	Rated/mean speed	-	2.0	-	1.73
12.	Rotor speed	rev/min	21.2	21.7	22.6
13.	Torque	MN m	1.78	4.4	2.96
14.	Platform		conc. grav.	TLP	conc. grav.
15.	Number of units		196	55	100pa
16.	Array efficiency	-	0.84	1	1
17.	Elec. trans. efficiency	-	0.9	0.9	1.0
18.	Availability	-	0.9	1.0	1.0
19.	System output	TWh/y	1.57	1.41	2.1
20.	System plant factor	-	0.245	0.293	0.342
21.	Capital cost/power	£/kW	1475	709	343
22.	Capital cost/rotor area	£/m2	1091	803	293
23.	Lifetime	y	20	-	25
24.	Ops and maintenance	p/kWh	0.5	0.55	0.3
25.	Onshore energy cost	p/kWh	5.6	5.0	1.6

TABLE 3 Cost estimates UK conditions
(HA machines, Vm = 9.5 m/s hub, 7.45% fcr)

			UK	US	Sweden
1.	Rotor diameter	m	80	106	102
2.	Rated power	MW	3.73	10	7
3.	System output	TWh/y	1.57	1.06	1.67
4.	System pl. fac.	-	0.245	0.221	0.273
5.	Capital cost	£/kW	1475	549	456
6.	Capital charge	p/kWh	5.1	2.1	1.4
7.	Maintenance	p/kWh	0.5	0.6	0.4
8.	Energy cost	p/kWh	5.6	2.7	1.8

TABLE 4 Energy cost estimates (p/kWh)
(100 m HA, Vm = 9.5 m/s, 20 y life)

UK	US	Sweden	Mean
4.18	2.84	1.83	2.95

TABLE 5 Target energy costs (at 5% dcf)

1.	Capital cost	500 £/kW
2.	Capital charge	1.7 p/kWh
3.	O and M	0.3 p/kWh
4.	Energy cost	2.0 p/kWh

TABLE 6 Target component costs (£/kW)

1.	Rotor complete	90
2.	Machinery	90
3.	Nacelle body	10
4.	Electrical (in nacelle)	30
5.	Steel tower (upper)	30
6.	Concrete tower (lower) and platform	80
7.	Assembly (in dock)	15
8.	Seabed preparation	15
9.	Installation (including float out)	10
10.	Grid connection (to shore line)	80
11.	Onshore grid connection	10
12.	Developer	50
13	Total	500

FIGURE 1

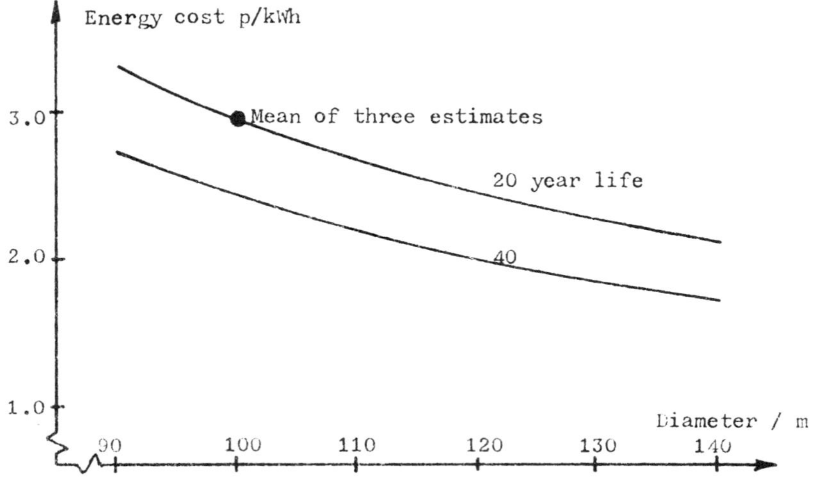

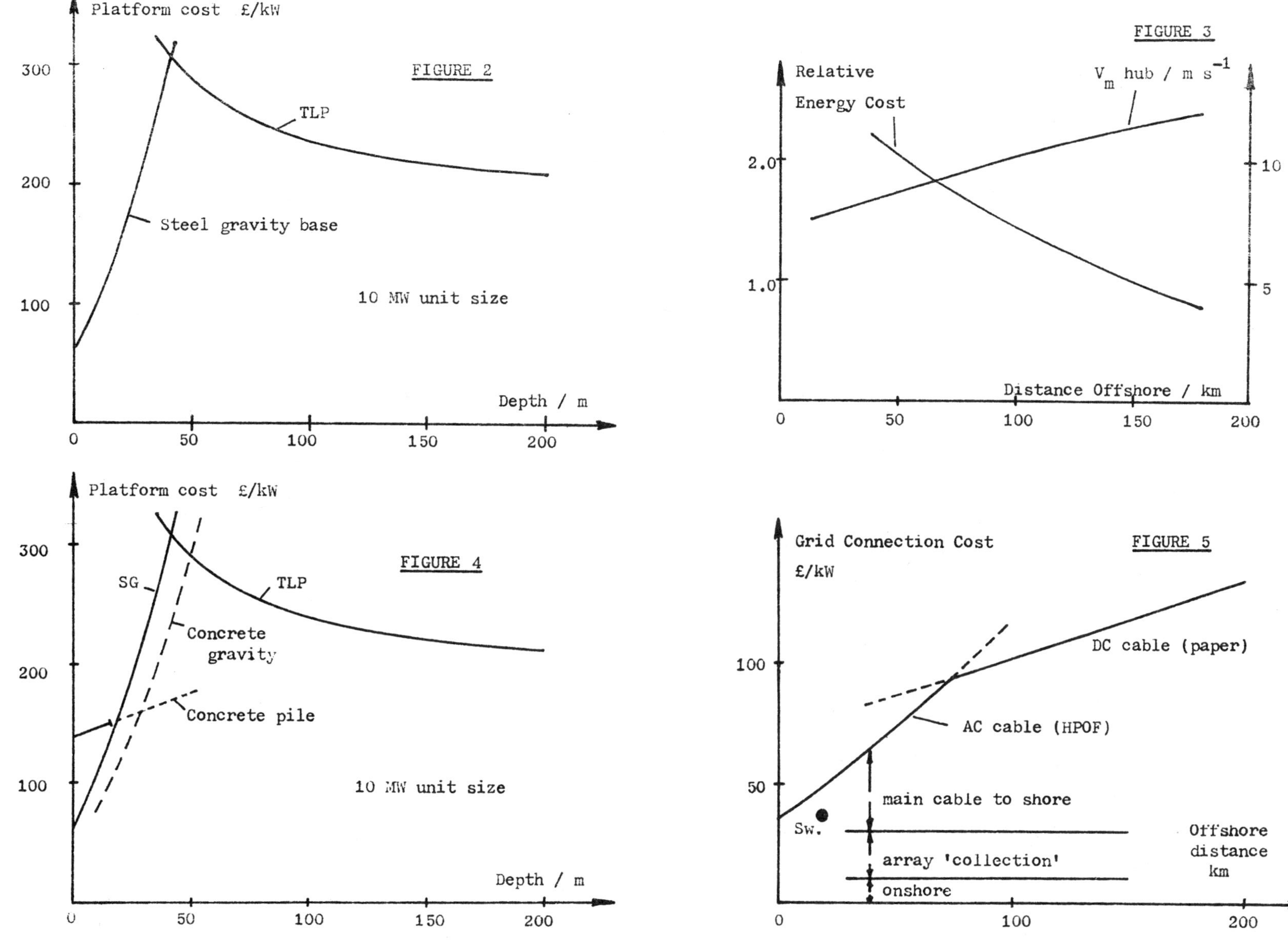
FIGURE 2
Platform cost £/kW
300
200
100
TLP
Steel gravity base
10 MW unit size
Depth / m
0
50
100
150
200
FIGURE 3
Relative
Energy Cost
V_m hub / m s^{-1}
2.0
1.0
10
5
Distance Offshore / km
0
50
100
150
200
FIGURE 4
Platform cost £/kW
300
200
100
SG
TLP
Concrete gravity
Concrete pile
10 MW unit size
Depth / m
0
50
100
150
200
FIGURE 5
Grid Connection Cost
£/kW
100
50
DC cable (paper)
AC cable (HPOF)
main cable to shore
array 'collection'
onshore
Sw.
Offshore distance km
0
100
200

EFFECTS OF WIND POWER AND PUMPED STORAGE IN AN ELECTRICITY GENERATING SYSTEM

G.E. Whittle
Department of Engineering
University of Reading

1. INTRODUCTION

The addition of wind energy (or any renewable, intermittent energy source) to an electricity utility increases the complexity of operating the system, since, in order to maintain security of supply it is necessary to compensate for the intermittent and largely unpredictable output from the wind turbines. This can be achieved by part loading fossil fuel steam turbine plant, or by increasing the use of expensive but rapid-response gas turbines, or if available by using pumped hydro storage plant. This is essentially the same procedure followed when dealing with demand and other system uncertainties, that is, wind power can be viewed as an additional uncertainty in the system. As a result when determining the fuel saving value of wind turbines, their effect on the system should be assessed and any operating penalty clearly identified.

Rockingham (Ref. 1) has suggested that for guidance a pessimistic estimate of this penalty can be arrived at by considering fossil fuel steam turbines to be part-loaded to the extent of the magnitude of the wind power input. This will result in a fuel penalty of about 13 to 15%.

Whittle et al (Ref. 2), using an electricity utility simulation model, have identified, for a 5300MW (12% of system maximum demand) wind turbine plant rated capacity, a minimum fuel cost penalty of 12%. This penalty was found to occur where fossil fuel steam turbine plant was part-loaded to 'cover' approximately 65% of the wind plant output, any further wind power uncertainty being serviced by gas turbines. This penalty assumed a ratio of two between the cost of gas turbine and steam plant fuel. Results for alternative ratios of fuel cost are presented in Ref. 3.

Based on work by Farmer et al (Ref. 4), and on their own work, Rockingham and Taylor (Ref. 5) have estimated a penalty of 5% where a combination of part loaded steam plant, gas turbines and pumped storage is available to service the wind power uncertainty.

In this paper the model described in Ref. 2 is used, in a more developed form, to investigate the use of wind power and pumped storage, and to assess operating penalties and fuel cost savings.

Capacity credit and value of wind power and pumped storage, obtainable from a reliability analysis of the system, have not been considered here.

2. DESCRIPTION OF THE MODEL

The modifications to the model are discussed below together with a review of its basic form.

The model is capable of considering the hour-by-hour operation of a number of types of generating plant in satisfying, over a full year, a realistic demand profile. It is appropriate to the Central Electricity Generating Board (CEGB) supply system, which has a maximum demand of 44GW and supplies 220 TWh of electricity per annum.

An estimate of the hourly system demand is made based on an interpolation of the CEGB annual maximum and minimum 24 hour load profiles. In order to represent demand uncertainty a random multiplication factor is generated for each hour such that the number is characterised by a mean of unity, a standard deviation of 0.015 and is generated from a normal distribution. This random number is used to multiply the estimate of the hourly load to produce a realistic degree of uncertainty.

The types of generating plant that can be represented in the model are nuclear, coal/oil steam turbines, gas turbines, pumped hydro storage and wind turbines. Nuclear plant is considered to be non load - following and operate on base load only. It is rated at 3832 MW and operated at a constant 75% load factor. Fossil fuel steam turbines are characterised by a start-up and shut-down time of eight hours. The specified unit rating is 454 MW which is representative of the mean size of power station on the CEGB grid. When operating, the fuel consumption is calculated assuming a full load efficiency of 35% (high merit) down to 30% (low merit) on a linear scale over 110 units. At part load the fuel use is determined as a linear variation with load factor, such that at zero load the fuel use is 15% of maximum. During start-up, a fuel use of 0.29 MW (thermal) per MW rated output is prescribed. When held at a steady state of readiness between cold and operational, a fuel penalty is imposed proportional to the state of readiness.

Both the pumped hydro storage and gas turbines are considered to be capable of instantaneous start-up in the context of the hour-by-hour model. In practice start-up to full load can be accomplished in less than 20 secs in the case of pumped storage (Dinorwic), and between 2 and 5 minutes for gas turbines. The pumped storage is rated at 2035 MW which represents existing plant plus the Dinorwic scheme. The energy storage level is specified as 10200 MWh, and the efficiency both of pumping and generating as 0.88. Gas turbines are considered to operate at a constant efficiency of 22.2%.

The calculated wind power output is provided from hourly values of wind speed from Scilly Isles for the year 1977, and a wind turbine characteristic appropriate to a multi-megawatt fixed pitch design. The mean wind speed for the site for the year considered was 8.6 m/s at a hub height of 45 m. The wind turbine used is rated at 1.77 MW at a wind speed of 17.2 m/s. The number of wind turbines at a site can be specified as input data to the model, for the present computations a rated capacity of 2000 MW has been specified.

When calculating the fuel cost of electricity supplied by fossil fuel plant, figures from Ref. 6 are used. For the case of coal/oil steam turbines a fuel cost of 0.431 p/kWh (thermal) is used, this is a mean figure for this type of plant in 1979/80 prices. It is assumed that the cost of gas turbine fuel is twice the above figure, i.e. 0.862 p/kWh.

No attempt has been made here to differentiate between coal and oil fired steam turbine stations.

System Operation

The system control logic of the model is as follows. Firstly, the constant steady output from the nuclear stations is subtracted from the system demand. Then the wind turbine output is subtracted up to the limit of the residual load. Any potential surplus of wind power is dealt with by notionally shutting-down a number of wind turbines.

Any residual demand is attempted to be met by operation of the fossil fuel steam turbines. Because of the start-up time a prediction of the load profile in the future eight hour time period is made at each hour. The decisions are then taken (again at each hour) regarding the start of the necessary numbers of units to meet the projected demand. Spinning reserve or spare capacity on the fossil fuel steam turbines operating, can be specified by treating it as a notional load on the system. Operating with spinning reserve incurs a fuel cost penalty because the steam turbines are then operated at less than full load and therefore at a reduced efficiency. Despite the penalty, this mode of operation allows some flexibility in servicing, at least partially, any demand and windpower uncertainty.

Any further residual load can be met by operation of pumped storage (if available), then gas turbines.

Pumped storage has been used in the model in two ways. Firstly, as a means of servicing system uncertainties where it is utilised in pumping mode whenever there is a surplus of windpower, that is, where nuclear plus wind power potentially exceeds system demand, or whenever there is spare capacity on the steam turbine units. The system is used in generating mode whenever the residual demand cannot be met by operation of the fossil fuel steam turbines; that is, pumped storage is used to service demand and wind power uncertainty. If operation of the pumped storage plant cannot satisfy demand then gas turbines are used.

The alternative way that pumped storage has been used is to combine the servicing of system uncertainties, as described above, with a peak-lopping operation. Peak-lopping is where the storage system is scheduled to meet demand at the time of daily peak and is scheduled for charging during the daily minimum demand. This has been arranged in the model by scheduling the fossil fuel steam plant to meet the residual maximum demand less the rated output of the storage plant at peak demand, and the residual minimum demand plus the pumping rating of the storage plant at minimum demand.

3. Discussion of Results

In order to assess the operation of the model in its revised form, it was used firstly to analyse the operation of a utility without wind power. Computations were carried out over a full years simulated period at a number of planned spinning reserve levels.

Figure 1 shows the variation in the fuel cost component of electricity supplied from the fossil fuel plant for a system without pumped storage, and for systems using storage for (a) meeting system uncertainties alone, and (b) for combined peak-lopping and servicing of system uncertainties. For the system without storage the minimum fuel cost occurs at a planned reserve of 0.01 of the hourly system demand. The increase in fuel cost at higher values of reserve is a result of the reduced efficiency of operation of the coal and oil fired steam turbines. Lower values of reserve implies a greater use of gas turbines, which have a lower efficiency and a higher fuel cost. Using storage in the system for servicing demand uncertainty reduces the fuel cost by displacing gas turbines. As can be seen the optimum reserve is reduced, and the fuel saving, approximately .006 p/kWh, corresponds to about £13m p.a. fuel saving based on 1979/80 fuel prices. If the storage is used in peak-lopping mode as well, then the fuel saving increases to £15 m p.a.

The variation in fuel cost with actual hourly mean values of spinning reserve is shown in Fig. 2. The optimum reserve levels for the three cases considered are 500 MW for the system without storage, 100 MW with storage used for servicing system uncertainty only, and 200-350 MW when storage is used for peak-lopping as well.

The use of pumped storage for peak-lopping saves fuel by displacing the use of low merit steam turbines (or gas turbines) which may be started and operated for only a short period of time to meet demand at the daily peak. Charging the system at the time of minimum demand utilises higher merit steam turbines which are already operating.

The maximum demand on the gas turbines is shown in Fig. 3 as a function of planned spinning reserve. For the system without storage the demand increases linearly with decreasing reserve, at optimum reserve (0.01) the gas turbine maximum demand is approximately 1500 MW. When storage is added then the gas turbine maximum demand at optimum reserve (-.005) is 1100 MW. In the case of additionally using storage for peak-lopping the maximum gas turbine demand at optimum reserve (zero) is shown to be 2000 MW. The gas turbine maximum demand is higher using the peak-lopping strategy since at the time of daily peak the storage has been scheduled for use in meeting predicted demand, reducing its ability to meet unpredicted peaks. The scatter of results in Fig. 3 is caused by the use of the randomised variable representing demand uncertainty (see section 2).

Figure 4 shows the annual output from the gas turbines for a system without storage. As expected the energy output increases as the planned spinning reserve is decreased. At the optimum value the gas turbine output is 183 GWh p.a., which when related to the maximum demand of 1500 MW results in an annual load factor 1.4%.

For a system where storage is used to service demand uncertainty only, the annual energy output from the storage and gas turbine plant, related to planned reserve can be seen in Fig. 5. At high values of spinning reserve there is little requirement for peaking plant, whilst at low values of reserve the storage is not being sufficiently charged and therefore is not able to deliver power when required. As a result increased demand is met by the gas turbines. At the optimum value of reserve the load factor of the storage plant, based on output, is approximately 5%. The maximum achievable load factor is related to pumping and generating cycle efficiency, and, for the system considered, is 39%.

The additional use of pumped storage for peak-lopping results in the annual energy output relationships illustrated in Fig. 6. At the optimum reserve the storage load factor is 9%.

Wind Power

For the system considered, the 2000 MW rated capacity wind component is expected to produce 5541 GWh p.a. This results in an annual load factor of 31.6%. Because of the fluctuating and, at times, unpredictable nature of wind energy the utilisation of wind power in the system can cause an operating penalty. This penalty is determined by comparing the fossil fuel cost both with and without wind energy, and relating the difference to the cost of providing the wind energy by fossil fuel plant.

This operating cost penalty has been determined for a range of planned spinning reserve levels for a system without storage, and also for a system where storage is used to meet demand and wind power uncertainties only. The planned reserve here is an addition of spinning reserve related to hourly system demand (chosen to be the optimum for the case without wind energy) and the reserve related to hourly wind power.

It can be seen from Fig. 7 that the minimum operating penalty for the system without storage is 5%, at a spinning reserve related to hourly wind power, of 0.4. This results in a fuel cost saving, for the 2000 MW wind system, of £79 m p.a. at 1979/80 prices.

The operating penalty is lower than previously found (Ref. 2), for two reasons. Firstly, the previous version of the model did not allow for any demand uncertainty, that is, the load was assumed to be fully predictable, resulting, for the system without wind, a higher plant efficiency. The addition of wind uncertainty into a system where uncertainty did not previously exist had a more marked effect. In other words the penalty of adding an uncertainty to a fully predictable system is greater than that of adding a second uncertainty (of similar magnitude). Secondly, because a range of steam turbine full load efficiencies have now been added to the model, the wind turbines now displace low merit steam plant.

For the system which includes storage (used for servicing system uncertainties only) the minimum penalty is -3%, that is, a gain of 3% over the apparent value of the wind energy. This corresponds to a fuel cost saving of £85 m p.a. at the planned spinning reserve level (related to hourly wind power) of zero. This shows that, provision of extra steam plant spinning reserve to compensate for the wind power uncertainty is unnecessary. The negative penalty, or gain, has resulted from a higher pumped storage load factor (cf. Figs. 5 and 9) which reduces spinning reserve on the system. Some of the benefit from this extra saving should be credited then to the pumped storage system.

Figure 8 shows the variation in gas turbine maximum demand related to planned spinning reserve. At the optimum reserve the demands are 2000 MW for the system without storage, and 2500 MW for the system with storage.

4. CONCLUSIONS

The effects on fuel usage of adding wind power and pumped storage to an electricity utility have been assessed using a computer based simulation model.

It has been found that a wind turbine system rating of 2000 MW could result in a fuel cost saving of £79 m p.a. at 1979/80 prices. Where pumped storage is also available the expected saving is £85 m p.a.

The operating penalty (if any) of using wind power is shown to be small.

ACKNOWLEDGMENTS

This work was carried out as part of a project to investigate the large-scale use of wind energy in the U.K., funded by the Science Research Council. The author wishes to acknowledge the contributions made to this paper by the other participants in the project, E.A. Bossanyi, P.D. Dunn, J. Halliday, N.H. Lipman and P.J. Musgrove.

REFERENCES

1. Rockingham, A.P. 'System Economics Theory for WECS' 2nd British Wind Energy Assoc. Workshop, Cranfield, 1980.

2. Whittle, G.E., Bossanyi, E.A., Maclean, C., Dunn, P.D., Lipman, N.H. and Musgrove, P.J., 'A Simulation Model of an Electricity Generating System Incorporating Wind Turbine Plant' Proc. 3rd Int. Symposium on Wind Energy Systems. Organised by BHRA, Cranfield, England, 1980.

3. Bossanyi, E.A., Whittle, G.E., Lipman, N.H. and Musgrove P.J., 'Wind Turbine Response and System Integration' 3rd Int. Conference on Future Energy Concepts. Organised by IEE, London, 1981.

4. Farmer, E.D., Newman, V.G. and Ashmole, P.H. 'The Economic and Operational Implications of a Complex of Wind-Driven Generators on a Power System' IEE Proceedings, Vol. 127 June 1980.

5. Rockingham, A.P. and Taylor R.H. 'The Value of Wind Turbines to Large Electricity Utilities' 3rd Int. Conference on Future Energy Concepts. Organised by IEE, London, 1981.

6. 'CEGB Statistical Yearbook' 1979-80. CEGB, London

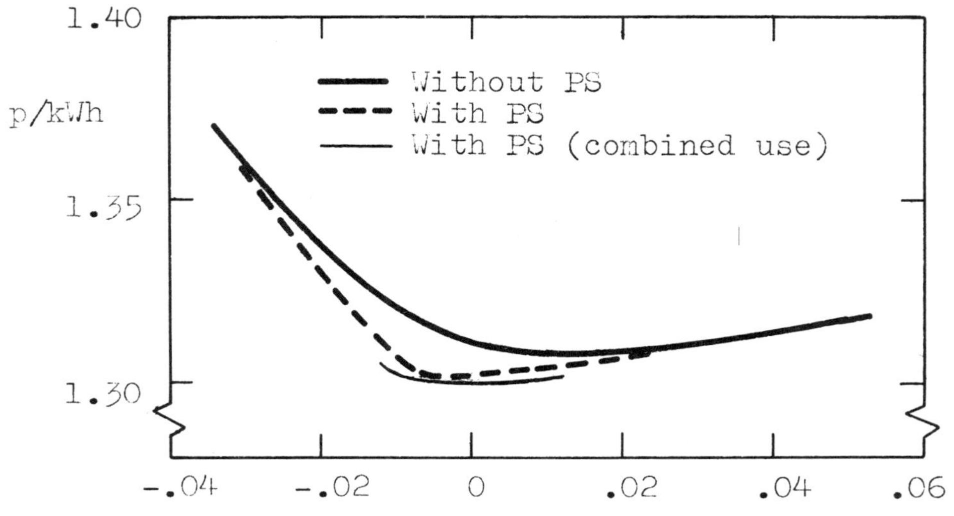

FIGURE 1. ELECTRICITY FUEL COST

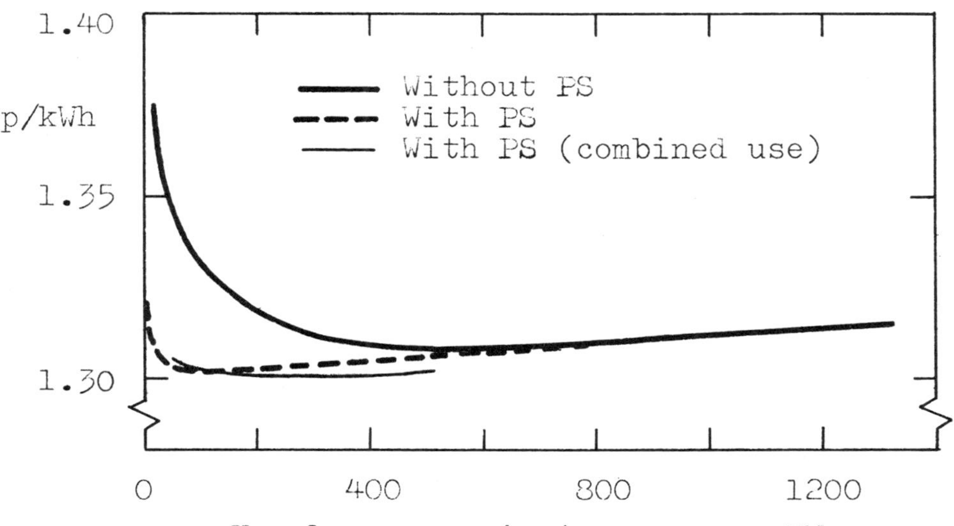

FIGURE 2. ELECTRICITY FUEL COST

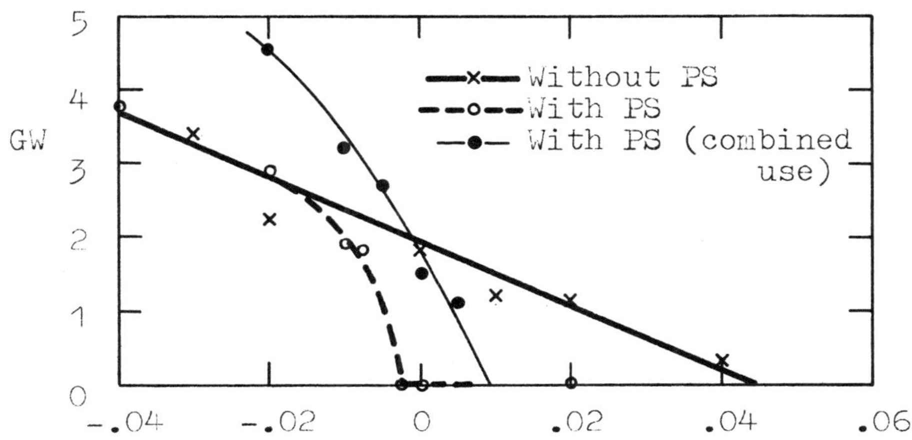

FIGURE 3. GAS TURBINE MAXIMUM DEMAND

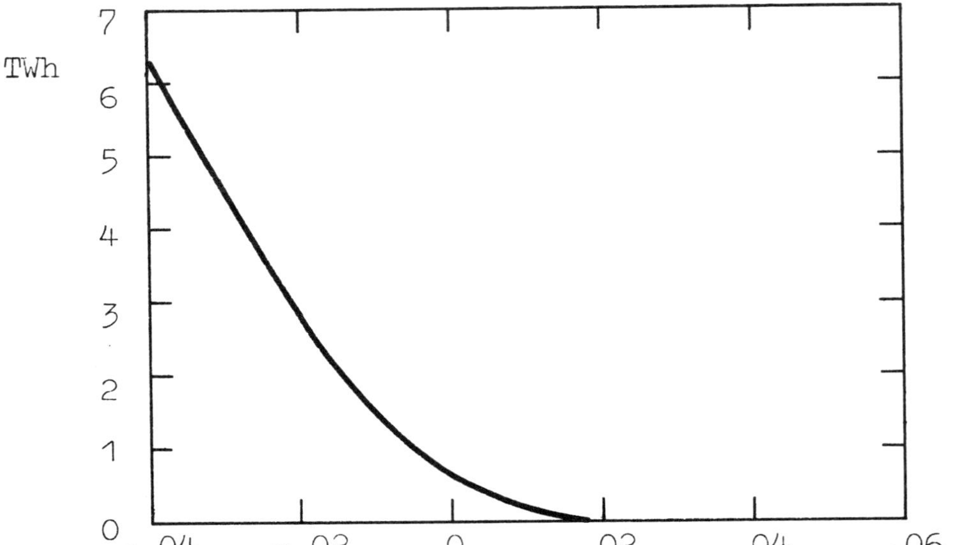

FIGURE 4. GAS TURBINE ANNUAL OUTPUT

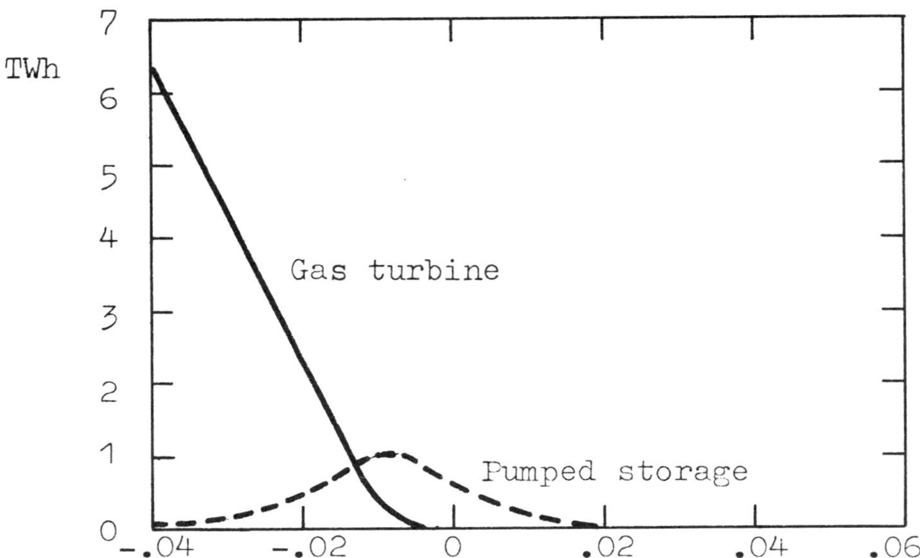

FIGURE 5. GAS TURBINE AND PUMPED STORAGE ANNUAL OUTPUT

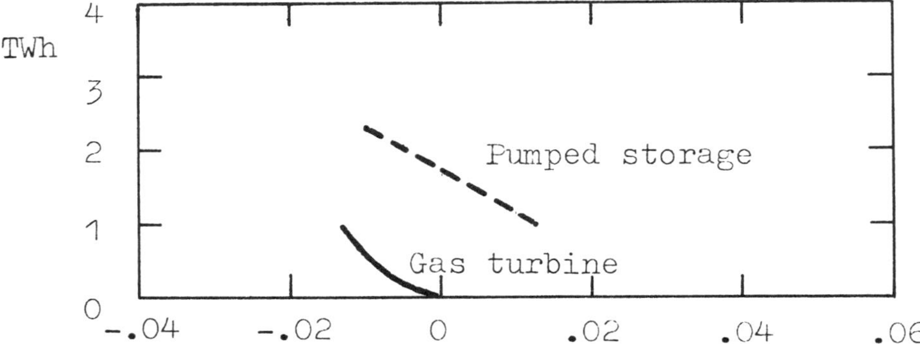

FIGURE 6. GAS TURBINE AND PUMPED STORAGE ANNUAL OUTPUT - 'PEAK LOPPING'

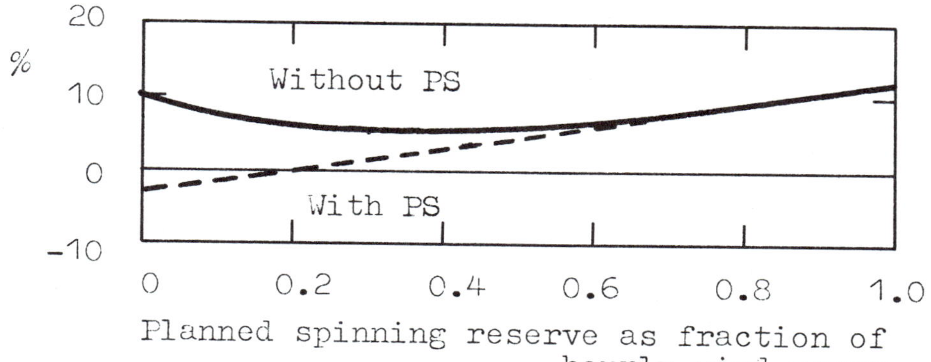

FIGURE 7. OPERATING PENALTY

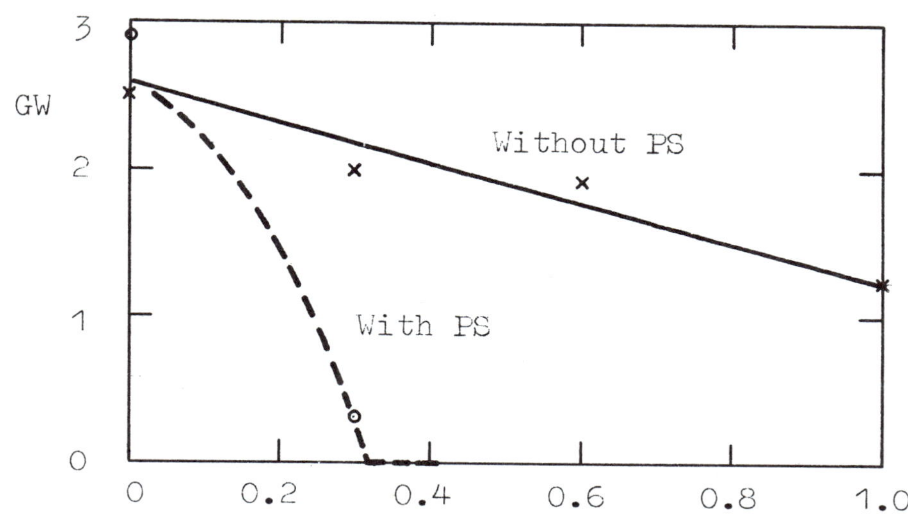

FIGURE 8. GAS TURBINE MAXIMUM DEMAND

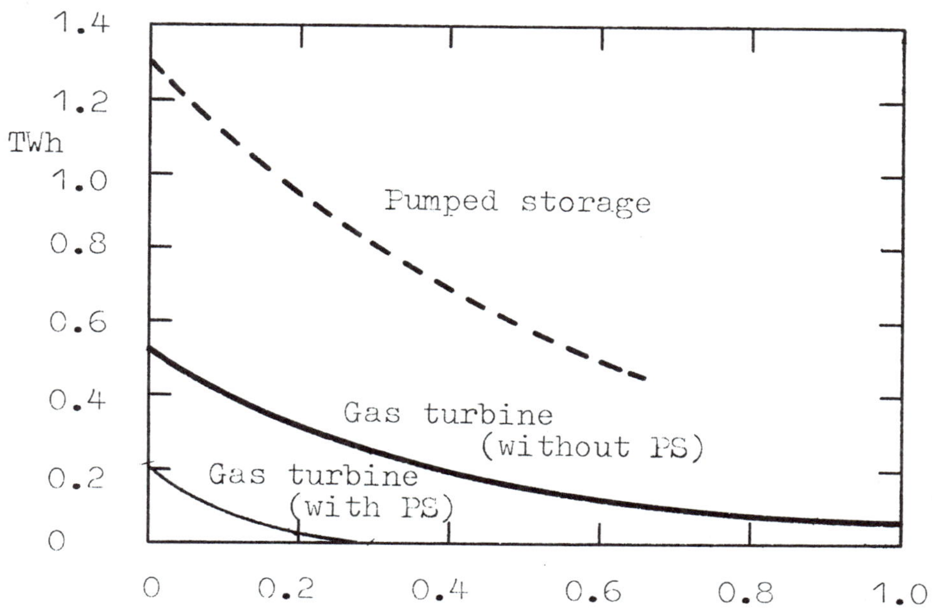

FIGURE 9. GAS TURBINE AND PUMPED STORAGE ANNUAL OUTPUT - WITH WIND POWER

THE FREQUENCY OF WIND TURBINE SHUT-DOWNS

E.A. Bossanyi, Reading University, U.K.

Abstract

Since a wind turbine must be switched off when the wind is either too high or too low, and because of the turbulent nature of the wind, a wind turbine may have to be switched on and off quite frequently when the wind speed is near either the cut-in speed or the furling speed. By making use of some one-minute wind speed data supplied by the Meteorological Office, this paper shows that one may have to expect several thousand shut-downs per year, and this may have implications for the life and maintenance requirements of the switchgear, and may also affect the turbine structure. The number of shut-downs can be reduced by making use of certain control strategies, which ensure that the turbine does not respond to short-lived wind fluctuations, but there may be a penalty in the form of a loss of energy output. Several different control strategies are investigated.

This paper summarises the main results of the study, which is reported in more detail in ref. 1.

Number of Starts near the Cut-in Speed

Figure 1 shows the number of starts per year occurring as a result of wind fluctuations around the cut-in speed. Since a whole year's data was not available, and also to generalise the results, the annual distribution of hourly mean wind speeds was assumed to follow a Weibull distribution (refs. 2, 3). The results are given for three different Weibull distributions, represented by the shape parameter, β. The figure also shows how the results change as the cut-in speed is varied relative to the site mean wind speed.

The upper set of curves, giving 10-20000 starts per year, is for a simple control strategy in which the wind turbine is turned on or off according to whether the average wind speed in the last minute was above or below the cut-in speed.

The other curves show that the number of starts can be considerably reduced by different control strategies. The different strategies are identified by four numbers: (x, N_1, N_2, M). A non-zero value of x implies a hysteresis strategy, in which the cut-in speed, originally the same as the cut-out speed, is increased by x%, leaving the cut-out speed unchanged. The wind turbine only cuts in if the average wind speed over the last N_1 minutes exceeds the cut-in speed, and it only cuts out if the average power output over the last N_2 minutes is negative. It is assumed that the machine motors if it is still connected when the wind drops below the cut-out speed, i.e. energy is supplied to the rotor to keep it turning at the right speed. M is the time it takes to turn the machine on from rest. Even a large machine can probably be brought up to operating speed in under a minute, but if a synchronous generator rather than an induction one is used then it may take some minutes to achieve synchronisation with the grid.

The energy loss associated with these strategies is fairly minimal (below 2%) as one is dealing with low power levels near the cut-in speed. The more extreme strategy with 100% hysteresis results in an annual energy loss of 5%. This was calculated using a wind turbine characteristic based on the 60 m. horizontal axis design described in ref. 4.

Frequency of Furling

The number of starts and stops around the furling speed is typically smaller than around the cut-in speed. However since the starts and stops are at full power the shock to the system is greater, and so is the effect on annual energy output. Figure 2 shows the annual number of furls as a function of furling speed, expressed relative to the annual mean wind speed for the site. Again the results are shown for three different Weibull distributions. The results are highly sensitive to both these parameters.

Control strategies at the furling speed are described by four numbers in a similar way to strategies at cut-in: (x, N_1, N_2, M). Now x% hysteresis means that the unfurling speed is x% below the furling speed, which remains unchanged. N_1 is the averaging time for unfurling, while the machine is furled if the mean wind speed over N_2 minutes exceeds the furling speed. M serves the same function as before. A wind turbine will probably be designed so that it must not operate above the furling speed for too long, and so only furling strategies with $N_2 = 1$ have been considered here. Figure 2 shows how increasing the hysteresis decreases the number of furls per year substantially (note the logarithmic scale).

Since furling occurs when the power level is high these different strategies can significantly affect the annual energy output. Figure 3 shows how the annual percentage energy loss increases as the number of furls is reduced by increasing the hysteresis. Several other types of strategy are also shown, but appear to be a little less effective than hysteresis. For these results the same turbine characteristic (ref. 4) was used, with the rated speed at twice the annual mean, giving a furling speed of around 2.5 times the mean.

Total Number of Shut-downs

As a further example, a wind turbine characteristic based on the Mod-2 wind turbine (ref. 5) and rated at 1.5 times the mean wind speed was used to produce figure 4. The histogram shows annual energy output above the line and total number of starts or shut-downs per year below the line, subdivided into those around the cut-in speed and those around the furling speed. This is shown for several different control strategies based on the Mod-2 specifications. Note the considerable loss of output with the 10-minute delay strategies (M = 10). Also shown is the option of increasing the furling speed, from 1.636 times rated (or 2.45 times the mean) to twice the rated wind speed (or 3 times the mean). This both reduces the number of furls and increases the annual energy output; however this strategy implies a higher turbine cost since it must withstand operation at higher wind speeds.

Conclusions

The results presented here have indicated that a wind turbine and its switchgear may have to cope with several thousand stops and starts per year. The number can be reduced by suitable control strategies, although around the furling speed this also results in a loss of energy output which can amount to 10% or more. This effect is much less significant around the cut-in speed. Increasing the furling speed is a desirable strategy as long as the cost of strengthening the turbine to operate in higher winds is not too great.

Acknowledgments

This work was carried out as part of a project to investigate the large-scale use of wind energy in the U.K., funded by the Science Research Council. The author wishes to acknowledge useful discussions with the other participants in the project, G.E. Whittle, P.J. Musgrove, N.H. Lipman and J. Halliday. Thanks are also due to D.J. Painting and the Meteorological Office for providing the one-minute wind speed data.

References

1. Bossanyi, E.A. — "Wind Turbines in a Turbulent Wind: Energy Output and the Frequency of Shut-downs". Wind Eng. (1981) 5, 12-28.

2. Swift-Hook, D.T. — "Describing Wind Data", RD/L/N 165/78, Central Electricity Research Labs. Leatherhead, 1979.

3. Bossanyi, E.A. et al, — "Wind Characteristics and the Output of Wind Turbines," Proc. 1st BWEA annual workshop, Cranfield, Multiscience, April 1979.

4. Dept. of Energy — "Development of Large Wind Turbine Generators", WPG 79/3, March 1979.

5. "2500 kW W.T.S. for Electric Power Generation" Mod-2 project Boeing Engineering and Construction, Seattle, Washington, Dec. 1978.

Figure 1. Shut-downs per year around the cut-in speed

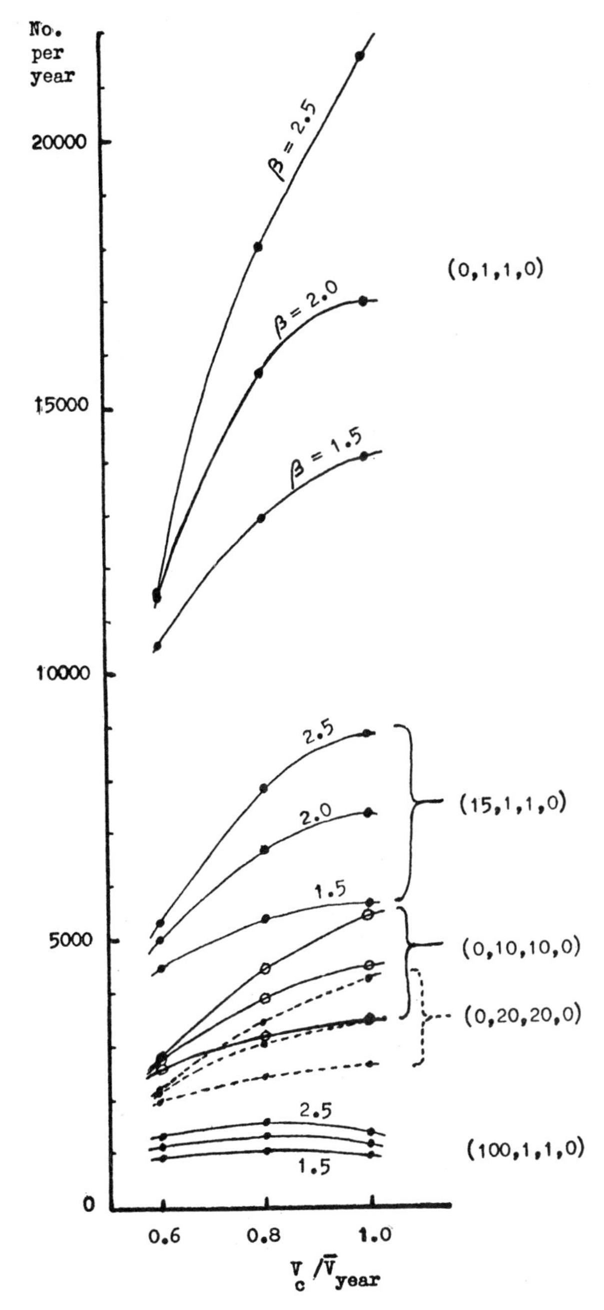

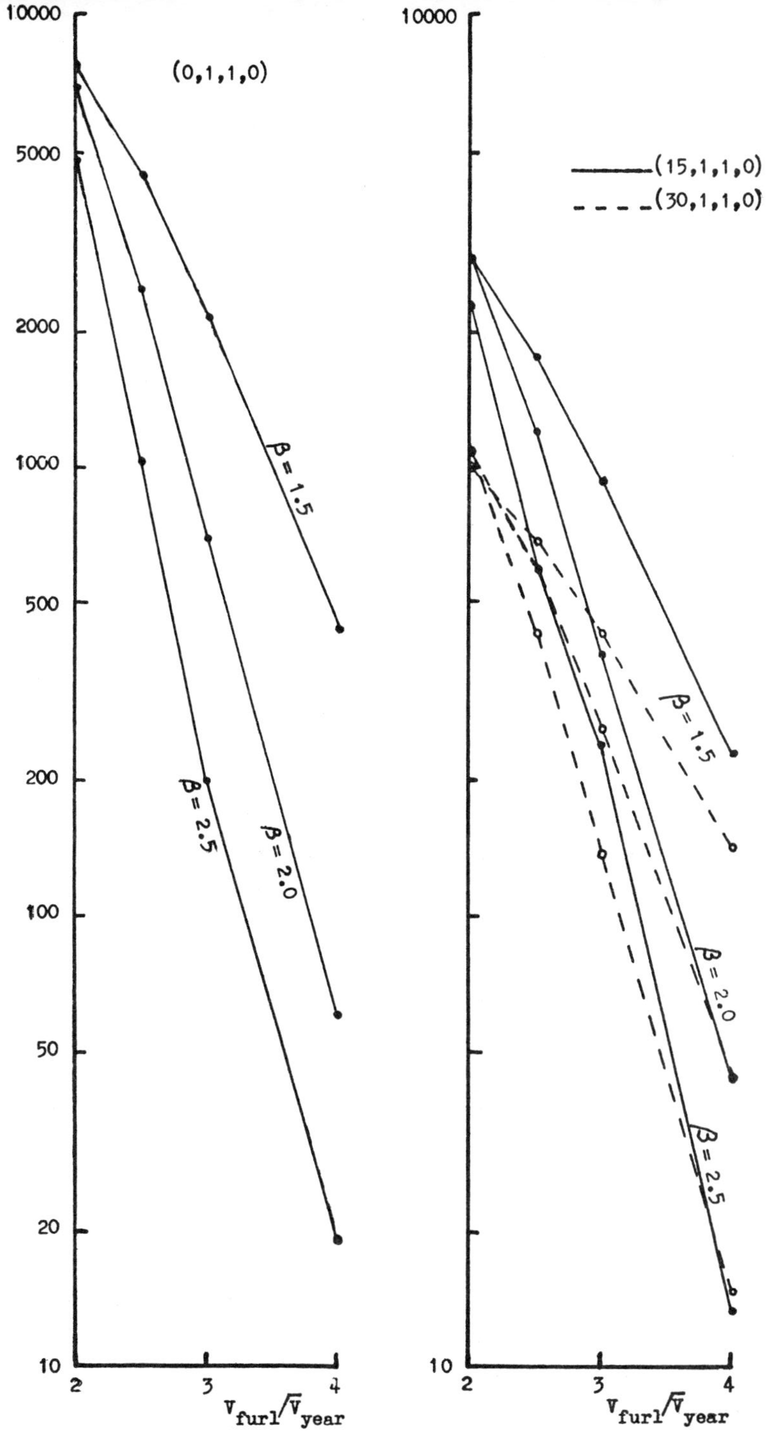

Figure 2. Shut-downs per year around the furling speed

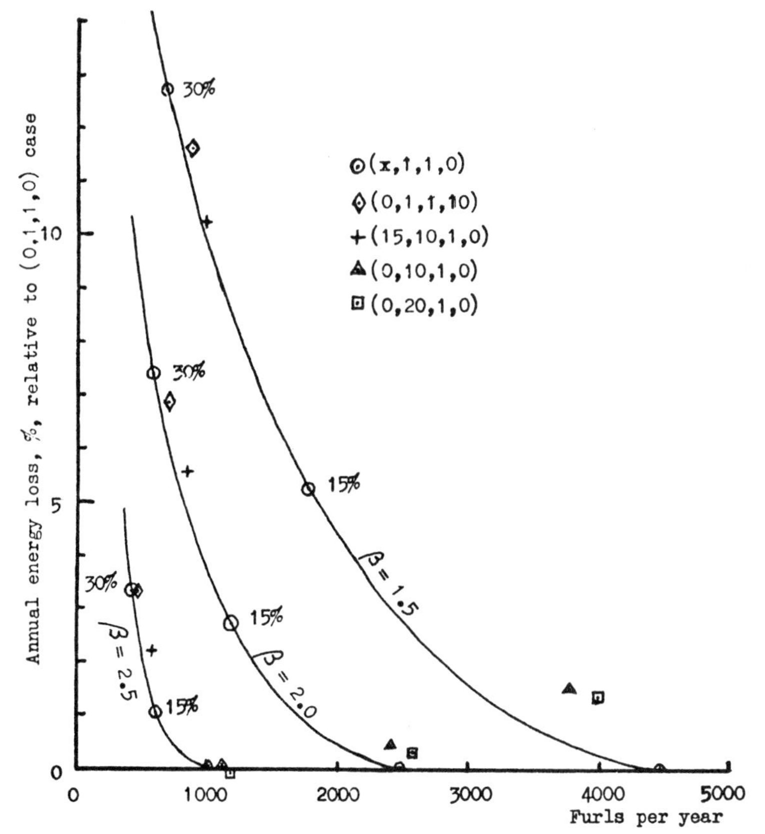

Figure 3. Energy lost with different unfurling strategies

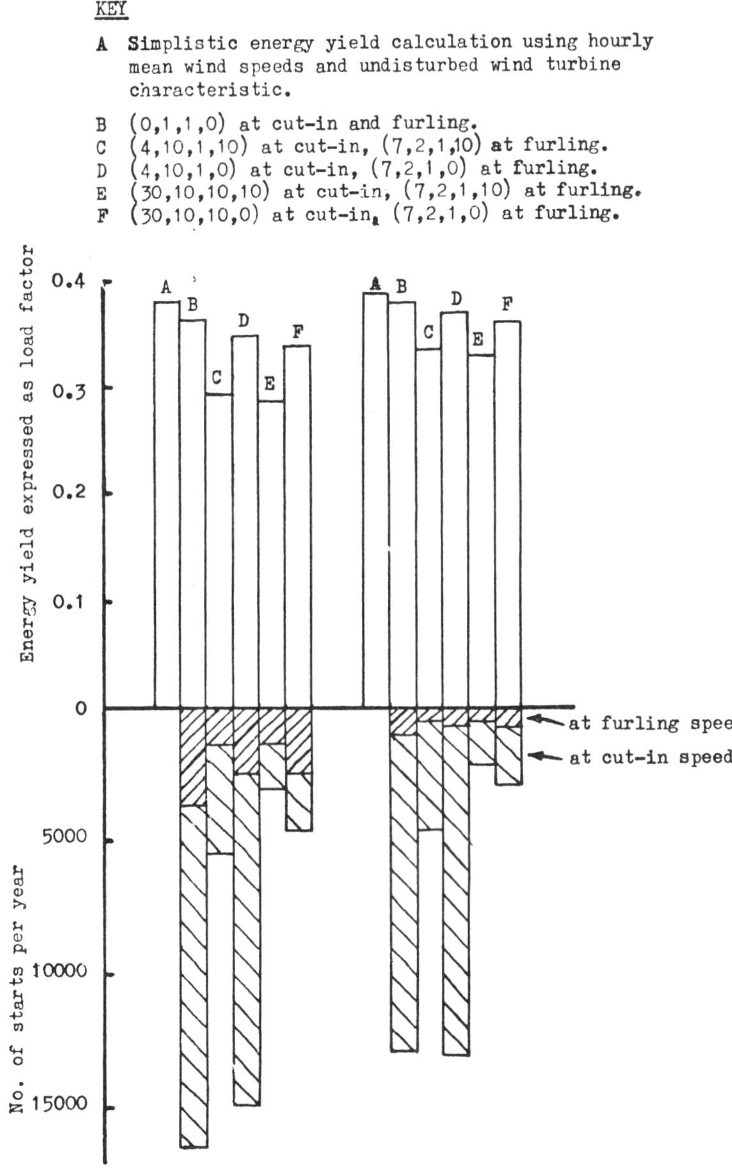

Figure 4. Energy and total shut-downs for Type 2 machine

A COMPARATIVE STUDY OF WIND POWER CONVERSION SYSTEMS

R.R. Wilson, P. Jamieson, A. Brown
James Howden and Co. Ltd.
J.R. Cure, J.A. Sullivan
Merz and McLellan

Abstract

James Howden and Company and Merz and McLellan have studied various systems designed to convert power from a wind turbine to electricity of voltage, frequency, and quality, acceptable to generating boards in the U.K. and elsewhere. A distinction is made between 'small' supply systems such as the island systems operated by the North of Scotland Hydro Electric Board (maximum demand 20/30 MW at winter peak) and 'large' supply systems such as main UK grid systems which are insensitive to local load and wattless power variation. Various types of wind power conversion system have been compared with regard to feasibility, cost, energy output, availability of components, and assessed as to their viability for small and large supply systems.

Introduction

The present analyses relate to machines with electrical outputs of several hundred kW. Only horizontal axis machines with fixed pitch blades are considered. Often the location of wind turbines, whether it be in offshore arrays, island communities, or isolated mainland sites, means that reliability and low maintenance are of the utmost importance. The use of complicated systems in an attempt to obtain the maximum energy output for a particular size of machine must be cost justified after proper account is taken of any extra maintenance needs.

Aerogenerators will cause voltage disturbances when connected to a public supply system. Starting will cause step type disturbances, and wind variations will cause ramp and flicker type disturbances. Voltage fluctuations may cause annoyance or even damage to equipment belonging to consumers connected to a public supply. Supply authorities are therefore concerned that the quality of supply should not be reduced to unacceptable levels by the connection of an aerogenerator. Figure 1 shows the voltage disturbance limits used by North American Power Utilities, and also recommended limits for the UK (1,2,3). The ratio of the output of the aerogenerator to the short circuit level of the system to which it is connected will principally determine the voltage fluctuations which occur.

In order that the voltage at the terminals of consumers in the vicinity of an aerogenerator does not experience unacceptable fluctuations or step changes, the output of an aerogenerator connected to a system possessing a low short circuit level must be of high quality, and not place large fluctuating reactive power demands on the system.

Large power supply systems possess such high short circuit levels that the load and reactive power variations due to an individual aerogenerator produce negligible effect on the voltage and frequency of the system as a whole.

Variable Speed Turbine with Rectifier Invertor System

When a fixed pitch turbine is used, the method of maintaining the aerodynamic efficiency at different wind speeds is to let the turbine speed vary with the wind to maintain a constant tip speed to wind speed ratio. Operation of the turbine at varying speeds allows the use of a simple untwisted blade. However, connection to a public constant frequency power supply requires an indirectly coupled drive system for the electric generator. Electrical decoupling between the aerogenerator and the electric supply may be obtained either by the use of a rectifier/invertor (figure 2) or slip energy recovery schemes using a slip-ring induction motor (4). There is little experience in operating this type of aerogenerator system. A rectifier/invertor system in itself is not new technology, but it is the integration of this into a complete aerogenerator system which has not been established. Several problems can be foreseen at this stage.

a) Rectifier/invertor schemes require a variable speed AC machine with excitation control and require complex voltage regulation circuits to enable inversion into the supply system.

b) Power conversion schemes using controlled rectifiers exhibit low efficiency at part loads due to rectifier volt drop characteristics and power requirements for control and cooling.

c) Reactive power variations and harmonic currents are fed into the public system.

In the U.K. the recommended limit of harmonic voltage distortion is defined by:

$$\sqrt{\sum (V_n)^2} < 2\% \text{ for } 11 \text{ kV}$$

where V_n = % of nth harmonic voltage magnitude.

Considerable problems may arise if rectifier/invertors are connected to rural distribution systems which possess low short circuit levels. Filter and reactive compensation equipment may be required to reduce harmonic voltage distortion and fluctuation and would increase the overall cost of an aerogenerator scheme considerably.

Variable Speed Turbine with Hydraulic Transmission

This system makes use of a hydraulic transmission system to transfer the power from the top of the tower to ground level (Figure 3). A large fixed displacement hydraulic pump is connected directly to the turbine shaft. The hydraulic fluid is fed to and from the nacelle via a rotary hydraulic coupling. At the base of the tower a variable displacement hydraulic motor is governed to run at constant speed and drive the synchronous generator. The use of a hydraulic system gives several advantages:-

a) The pumps can be directly driven from the turbine shaft. If the displacement of the motor is smaller than that of the pump, this will give a corresponding increase in speed, no gearing will be required.

b) The only items at the top of the tower will be the turbine itself and the pump, all other equipment can be mounted at ground level.

c) The hydraulic system can give a high degree of control, it is possible to obtain a high quality output, both by smoothing out variations in torque, and by allowing the use of a synchronous generator.

d) Although the peak efficiency of the hydraulic system will be lower than many alternative transmission methods, the drop in efficiency at part loads is much less, giving a high overall energy output.

The main problem in using a hydraulic system would be in obtaining suitable equipment. A hydraulic system of 30 kW capacity has been tried on the Aldborough aerogenerator (5). With directly coupled pumps and motors, 200 kW is probably the maximum size that could be put together with a good system life at present. The maintenance-free life of large hydraulic units may be five years or less. One solution is to use a gearbox and couple multiple pumps and motors into the system, as in the Bendix 3 MW machine. Equipment is under development in this country for hydraulic marine propulsion systems of megawatt sizes but it is several years from commercial production (6).

Variable Speed Turbine - With Scoop Controlled Variable Speed Fluid Coupling

This system utilises a variable speed, scoop controlled fluid coupling, mounted on the output shaft of the gearbox, to allow the generator to operate at a constant speed while maintaining the high efficiency of a variable speed turbine. Variable speed hydraulic couplings are a well proven product, having been used extensively in the drives of fans and pumps at powers in excess of those considered for aerogenerators. However in a wind turbine system, the coupling does not operate in the normal way. The greatest slip is required when handling the peak power, and the efficiency therefore falls as the turbine speed increases. A cooler is thus required, and the overall power output is reduced.

Constant Speed Turbine with Induction Generator

It is possible to operate a fixed pitch turbine at constant speed in varying winds, and by using twisted blades, a satisfactory efficiency can be obtained. Consequently the blades are often more expensive than those used in a variable speed turbine. Induction machines operating in the generating mode operate at low power factors. If the supply network is unable to supply the necessary reactive power for normal operation, it is usually found that synchronous generation is more cost effective than the installation of capacitors for power factor correction.

During wind gusts the torque variations appearing at the generating shaft depend primarily on the aerodynamic characteristics of the aerogenerator blade system. Induction machines possess a high pull-out torque when operating in the generating mode, which is usually 50% greater than in the motoring mode, however this pull-out torque is only obtained if the machine terminal voltage is maintained under dynamic conditions. Transient torque variations in the aerogenerator drive system may be reduced during starting and running conditions by the use of a fluid coupling between the generator drive shaft and the gearbox.

Real and reactive power cannot be controlled by induction machines, and power pulsations, due to wind speed variations, are fed into the public supply. If the voltage fluctuations induced in the public supply are outside recommended limits, it may be more cost effective to reinforce the supply system rather than attempt modifications to the aerogenerator blades or drive system (7,8).

The use of constant speed aerogenerators driving induction machines provides a simple and robust system, and provided that the size of the aerogenerator is small in relation to the short circuit level at the point of common coupling to the supply system is generally satisfactory (9).

Constant Speed Turbine with Synchronous Generator

This system is shown in figure 5. It is very simple mechanically, giving obvious advantages in terms of reliability and low maintenance. Synchronous generator systems can if required incorporate some kind of soft shaft arrangement (10), such as a fluid coupling to reduce disturbance to the public supply system.

Synchronisation with the public supply may be accomplished by three methods

a) fine control of turbine speed by aerofoil trimming

b) application of field excitation while running at slip speed to pull the machine into synchronisation with the supply system

c) the use of a load bank to provide load control in fine steps, and thus speed control.

Of the three methods, method (b) is the most positive and could be used under varying wind speed conditions. Method (a) would cause little voltage disturbance to the supply system at the instant of synchronisation, but synchronisation may not be possible under variable wind speed conditions. Method (c) is most suited to weak supply systems (11).

Voltage variations in the public supply can be minimised by the use of excitation control. The pull out torque of the machine can also be increased by obtaining a control signal from the supply system voltage transformer.

Synchronous machines, unlike induction machines, are stiff against the supply system, and would oscillate at a natural electro-mechanical swing frequency determined by the aerogenerator inertia and tie line reactance, unless the machine is adequately damped. Wind power fluctuations, caused by the variation in wind speed in the vicinity of the tower, produce cyclic torque variations, and may sustain machine electro-mechanical oscillations. The oscillation of the generator rotor may be damped by increasing the effective rotor damping by excitation control (12).

Multi-Speed Systems

A method of obtaining a high efficiency with a fixed speed turbine without the problems of a variable speed transmission is to operate the turbine at a number of fixed speeds. This could be done by means of various gear ratios or more likely by the use of pole switching in the generator.

The feasibility of this system has been investigated (13), and the economic advantage evaluated. It will produce more power for a small additional capital cost. However speed changing with induction machines involves large reactive power fluctuations. This may be acceptable on very large systems, but could cause severe system disturbances on a weak system. Some decrease in reliability would be expected compared with a single speed system. Further work would be required to arrive at a commercial system.

System Power Output and Cost

The relative power output for a full year for each system was evaluated for a high energy site typical of the Scottish West Coast. To determine cost effectiveness of each system it was convenient to consider a turbine diameter of 18m with a maximum rated power of approx. 200 kW and compare costs and annual energy outputs for each system. More rigorously, a required annual output could have been fixed and turbines sized and rated (in general each differently) to meet this most economically. However within the accuracy to which efficiences, reliability (hence maintenance costs) and capital costs can be estimated, it scarcely matters which approach is adopted. Final costs were based on the complete set of components for each system. Cost of the tower was constant. Costs for gearboxes, generators etc. were based on manufacturers quotations. A simple symmetrical blade was costed and costs adjusted where extra complexities, twist or tip spoilers etc. were required.

The relative costs are presented in table 1 for the complete aerogenerator, including tower, foundations and instrumentation. It can be seen that the most cost effective systems are the variable speed turbine with hydraulic transmission and the constant speed turbine with synchronous generator. The cost of an induction generator system would be very similar to that for a synchronous machine.

Conclusions

Rectifier/invertor systems were found to be more costly than the other schemes examined, and in addition may cause deterioration in the quality of public supply, particularly if connected to weak rural supply systems.

Hydraulic power conversion systems allow the use of a conventional AC generator coupled directly to the supply. Torque variations due to wind speed variations can be smoothed out using hydraulic control. However, it has yet to be established if hydraulic components will have sufficient life for this application.

The use of aerogenerators utilising induction machines is generally satisfactory provided that the size of the aerogenerator is small in relation to the short circuit level at the point of common coupling to the public supply system.

Synchronous generators are well suited for connection to supply systems possessing low short circuit levels, since they exercise dynamic control over system voltage and are able to supply reactive power. This system appears to have the lowest initial cost and should prove reliable.

Acknowledgements

The authors wish to thank the directors of James Howden & Company Limited and the partners of Merz and McLellan for permission to publish the information contained in this paper.

References

1. Chief Engineer's Recommendations - U.K. Electricity Council. (a) P7/2, (b) P8,(c) P9, (d) P13, (e) G5/3.

2. British Electricity Board Reports - Chief Engineer's Conference. (a) No. 15, (b) No. 26, (c) No. 73.

3. Electric Utility Flicker Limitation M.K. Walker IEEE transactions on industrial applications Vol. A15. Nov, Dec. 1979.

4. Electrical Equipment for a Large Wind - Power Plant. H. Muhlocker - Siemens Power Engineering, Feb. 1980.

5. The Design and Construction of the Aldborough Aerogenerator. W.R. Nickols Proc. IEE Conference Future Energy Concepts, London 1979.

6. Wind Power Generation for National and Regional Grid Systems; Design Proposals for Generation and Transmission R. Freer. Third International Symposium on Wind Energy Systems, Copenhagen 1980.

7. Danish Experience of Mains Connected Private Aerogenerators. C.J. Daniels - Distribution Development December 1980, Issue 80/4.

8. General Guidelines for the Connection of Small Private Windmills to Low Voltage Networks. Danish Association of Electricity Works (DEF) August 1976/Electricity Council Translation June 1980.

9. Development of Large Wind Turbine Generators. Department of Energy. WPG 79/3 March 1979.

10. Drive Train Assembly of the Swedish WTS 3. N.G. Byggeth, K.E. Halsten, L. Thoreson. Third International Symposium on Wind Energy Systems, Copenhagen 1980.

11. EPR1/DOE Workshop Presentation : MP1-200 Control System Design. WTG Energy Systems Inc. March 79, Monterey, California.

12. Techniques for Stabilizing Wind Turbine Generator connected to Power Systems. H.H. Guo T. Hwang, H.V. Monzeicog. IEEE. PES Meeting, January 1978, Paper A78 280-0.

13. Integration of Wind Power onto an Electricity Supply System. R.H. Taylor, R.J. Leicester, G.E. Gardner, P.J. Franklin. Proc. of 1st B.W.E.A. Wind Energy Workshop, 1979.

System	Relative Cost	Relative Energy output	Relative Cost per Energy Unit
Variable speed rectifier/invertor	1.30	1.20	1.08
Variable speed hydraulic system	1.22	1.24	0.97
Variable speed fluid coupling	1.05	0.76	1.38
Constant speed synchronous gen	1.00	1.00	1.00

TABLE 1 - RELATIVE COSTS

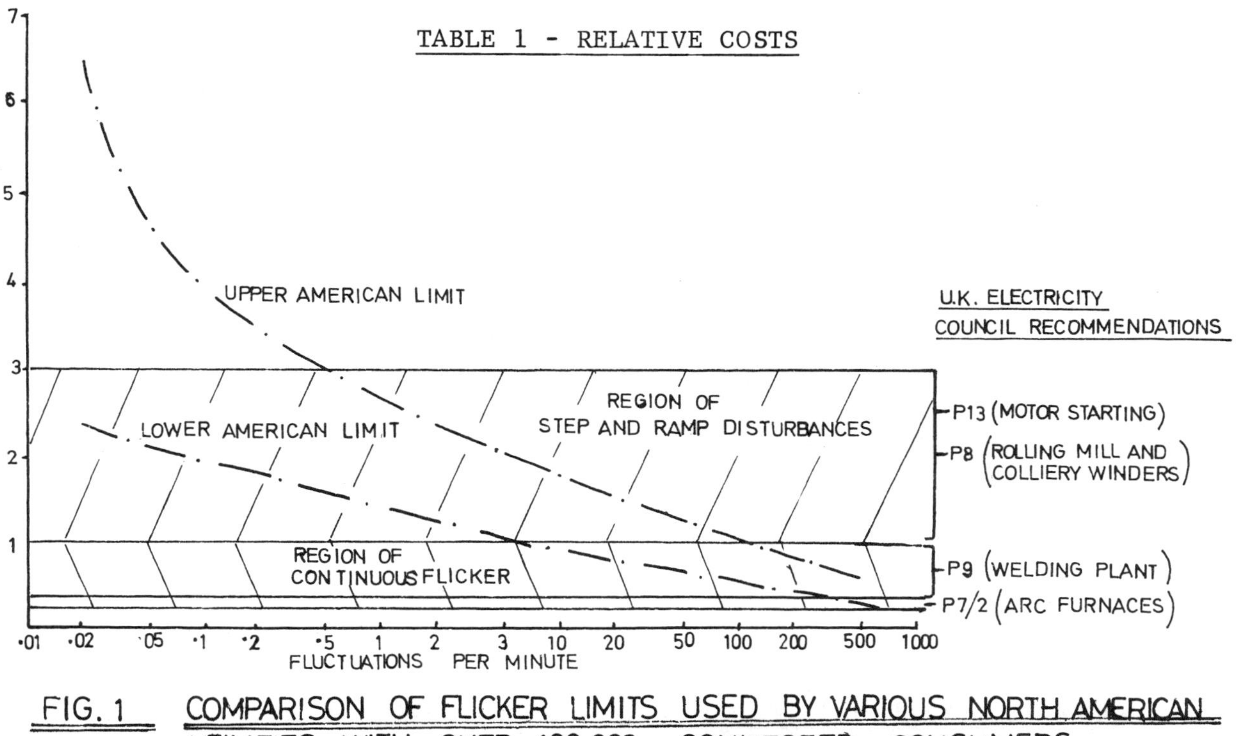

FIG. 1 COMPARISON OF FLICKER LIMITS USED BY VARIOUS NORTH AMERICAN UTILITIES WITH OVER 100,000 CONNECTED CONSUMERS

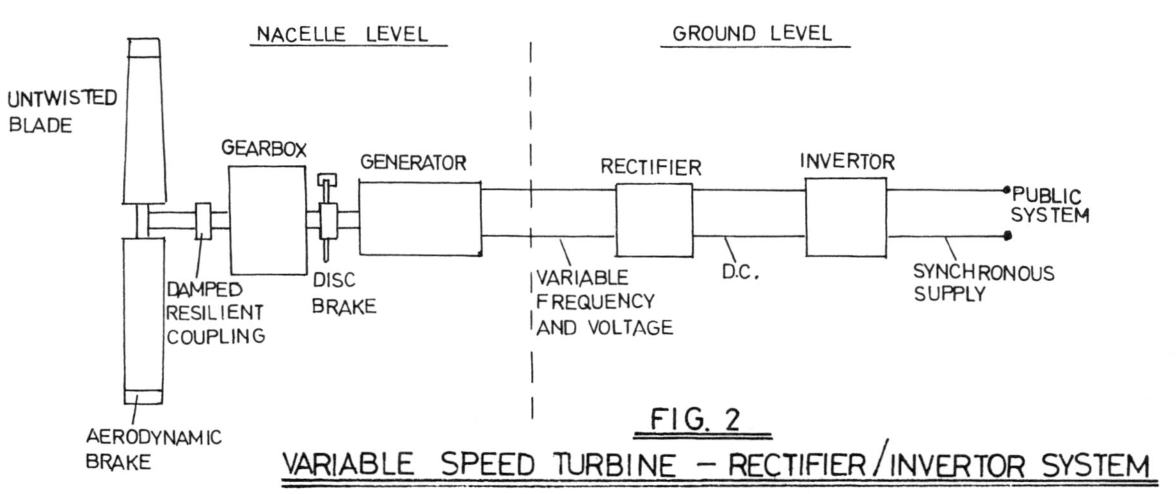

FIG. 2

VARIABLE SPEED TURBINE - RECTIFIER/INVERTOR SYSTEM

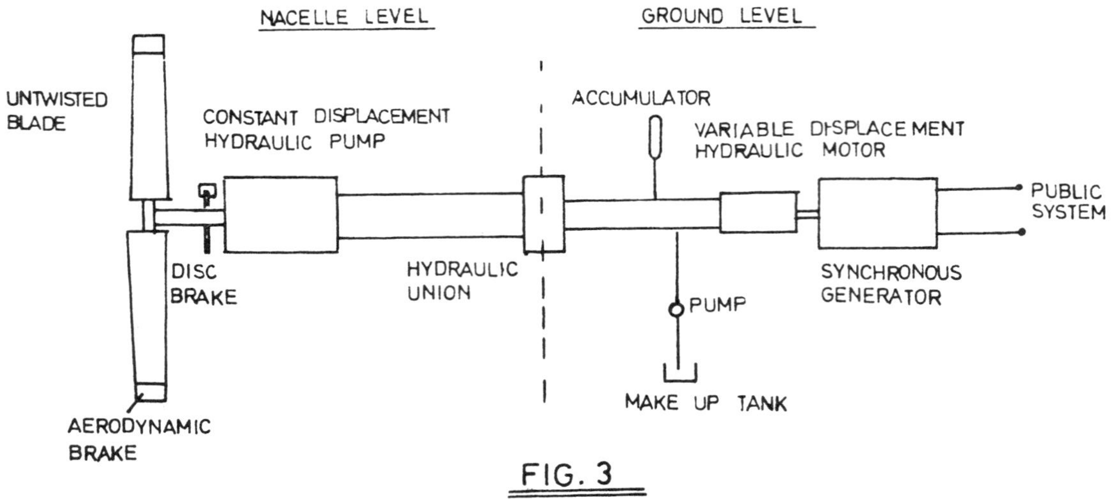

FIG. 3

VARIABLE SPEED TURBINE - HYDRAULIC TRANSMISSION SYSTEM

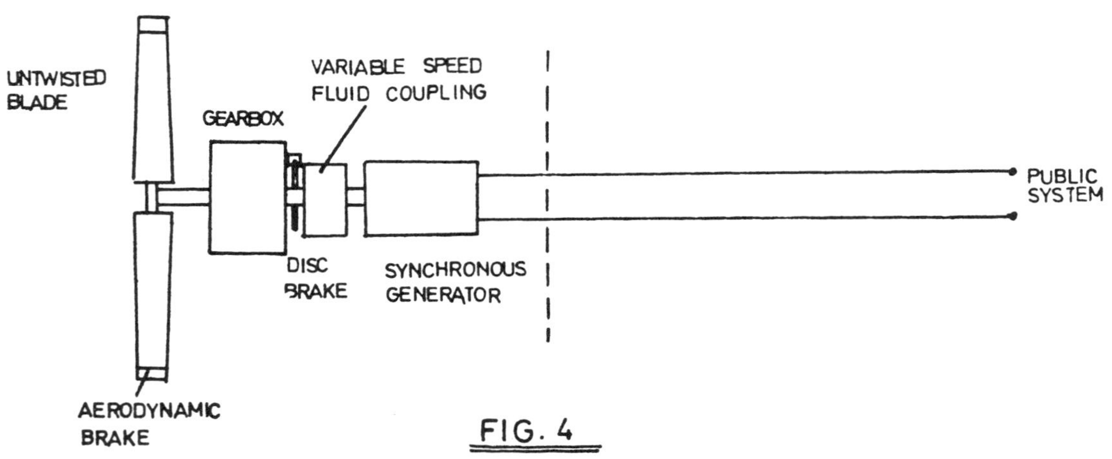

FIG. 4

VARIABLE SPEED TURBINE - VARIABLE SPEED FLUID COUPLING TRANSMISSION

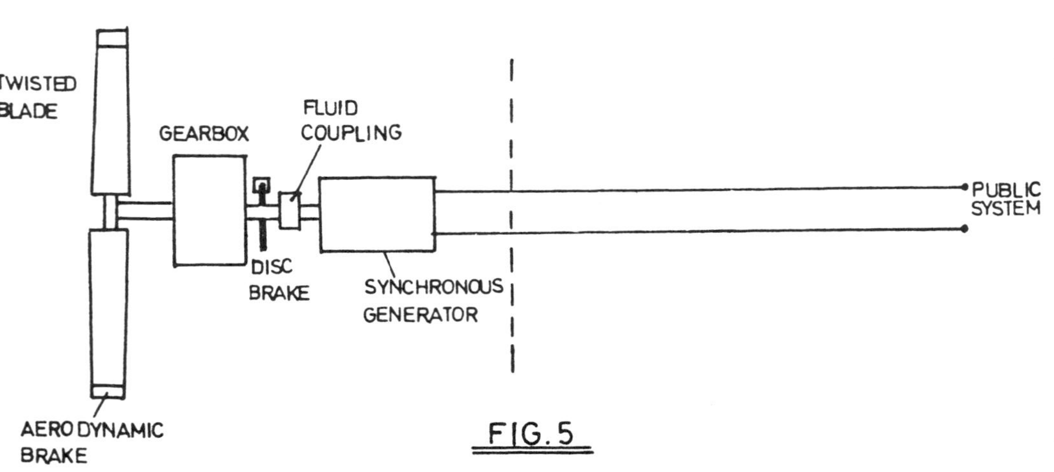

FIG. 5

CONSTANT SPEED TURBINE - SYNCHRONOUS GENERATOR

ENERGY REQUIREMENTS OF LARGE WIND TURBINE SYSTEMS

J C Dixon and R J Lowe
Open University, UK.

Abstract

A full process analysis of large land and offshore wind energy systems is performed. The land system is found to have a payback period of 0.47 years and an energy gain of 50. The offshore system in shallow water (20 m) has a payback period of 0.37 years and an energy gain of 55. These figures appear to be superior to those exhibited by nuclear systems, and therefore it appears that as a fuel-saving technology, from the energy analysis aspect, wind power is very competitive.

Introduction

This paper gives a brief summary of an investigation of energy inputs to wind energy systems on land and offshore (Ref 1). It is restricted to the quantitative results, rather than comments on the implications. The major energy inputs to large wind turbine systems are through the materials used; primarily steel and concrete, but potentially including aluminium alloys, copper, lead, GFRP, CFRP and others. There are energy inputs for construction, both direct (fuel used by cranes, tugs etc) and indirect (energy costs of building cranes, tugs etc). There are also energy costs of building and running support facilities, and providing operational supervision and maintenance. Minor items have been found to have a significant cumulative effect.

Conventions

An energy process analysis gives an estimate of the energy requirement of a product, according to the energy content of the input materials and components and of the processing costs. We have generally adopted published values for wind turbine components in the most complete form for which reasonable data is available. For example, a published value for machined steel equipment is used for the gearbox. On the other hand, we know of no published value for marine power cables and so have performed a process analysis with inputs such as steel wire and copper wire. Energy requirements for the various inputs used in this paper come from a number of sources: steel and steel components (2), aluminium (5,6), electrical equipment (7,8), road construction (9), transport (10), general construction on land (10,11).

One difficulty in estimating the energy requirement of a metal is that it is less for recycled material than for new material from the ore. (A particularly extreme case is aluminium; 30 MJ/kg against 300 MJ/kg.) Strictly the value used should depend upon the marginal fraction of recycling (12), purity requirements and so on. Such information is not available, so we have followed the usual convention of adopting the overall mean value for the UK. (Hence aluminium is 110 MJ/kg.)

Ready-to-use material such as sheet or wire is significantly more energy intensive than the ingot material. This depends upon the energy efficiency of the industry (Japanese steel averages less energy than British) and the individual plant (modern plant is more efficient than old). Such problems closely parallel economic assessment difficulties. We have generally followed the energy accounting conventions of (13). We estimate that the energy cost of demolition and disposal is quite small and will be approximately equal to the energy credit of recyclable material, and hence these are neglected.

The electrical energy yield of the baseline design analysed is 40 TJ/year. Since the role of wind power is anticipated to be as a fossil fuel saver this will save primary thermal energy at the rate of about 120 TJ/year (the precise thermal/electric ratio actually depends on the efficiency of the offset plant and on the requirement for spinning reserve etc). The payback time periods listed in the tables are based on this value. In assessing offshore system requirements, to include allowance for appropriate onshore construction and maintenance facilities and marine cable production, it is assumed that a total of about 3 GW is installed over a period of ten years (72 machines per year).

Land Systems

The baseline design analysed (figure 1a and Table 1) is a teetered-rotor, flexible-tower horizontal-axis type exemplified by modern developments in the American wind energy programme (14,15,16,17). The baseline array (Table 2) totals 400 MW. Table 3 shows the system breakdown adopted. Most design features are derived from averages of published values, for example the above-tower-masses of figure 2. This is subdivided into blades (GFRP), machined steel, fabricated steel and electrical components (Tables 3 and 4), since energy requirements data are available under such headings. The tower design (Table 5) is of the compliant steel shell type (figure 1a, refs 15,16) with an adequate foundation for soft ground (Table 6). The baseline machine is summarised in Table 7. The total of about 28 TJ, however, does not yet include various array and construction costs.

The implications of using various rotor materials are shown in Table 8. Some special difficulties here include estimation of the material masses, which will depend on the rotor type, and the specific energy requirements of materials such as aluminium. The value of 4000 GJ/t for carbon fibre reinforced plastic (18) has a drastic effect, although this may be justified by the good fatigue properties of CFRP. There are several possible tower alternatives. An appropriate rigid steel lattice tower appears to mass some 320 t according to Refs 14 and 15, requiring an extra 6 TJ (0.05 years). However Ref 19 claims a rigid lattice tower of much lower mass per unit installed power.

A rigid or compliant tower could be built from reinforced concrete. The mass of such towers will be much greater than pure steel ones. If concrete is used only in the lower tower, and if the design case is overall tower rigidity, then the required cross-section of concrete will depend on the modular ratio of the materials (ie E_c/E_s). For a strong concrete (typically E_c = 32 GN/m2, ρ = 2400 kg/m3) the required area ratio is 6.6 and mass ratio 2.0. The specific energy requirement of concrete is typically 1.5 GJ/t, against 30 GJ/t or more for steel, so it appears that the energy requirement will be greatest for the all-steel tower (even though there may be design limitations due to wall thickness of concrete, and there is a lower limit on the amount of steel because of the need for reinforcing and prestressing, figure 3).

If concrete is also used in the upper tower then a further effect must be considered. The extra concrete mass adds to the effective inertia, which increases the required stiffness for a given frequency. The steel to concrete replacement mass ratio is thereby increased from R to R', figure 4, where a is the fraction of tower mass effective as inertia at the tower top, m_1 is the tower mass and m_0 is the nacelle plus rotor mass (a = 0.16 for the baseline design). The steel to concrete ratio is still much less than their energy requirement ratio. It therefore seems likely that the all-steel tower is the conservative (ie worst) case from the energy analysis standpoint.

Table 9 shows the complete land-system energy requirement (ie the baseline machine installed complete with all the array electrical connections). Construction includes: transport of all subassemblies, cement and aggregate and spoil removal totalling 0.80 TJ (160 Mt m at 5 MJ/t km); site work, excavation, craneage, concrete mixing and pumping, electricity, lighting etc totalling 0.20 TJ; access roads 1 km per machine, 4 m wide at 1.5 GJ/m^2 giving 6 TJ. The large value for access roads is interesting. It is probably rather conservative since no useful existing roads are assumed. The array electrical item includes 6 kV underground cable (1.66 TJ), transformers (0.60 TJ), switchgear etc (0.20 TJ).

The complete land system value of 37.3 TJ is 34% higher than the bare machine parts of Table 7. It corresponds to a payback period of 0.31 years at the rather high energy yield (for the flat land machine) of 40 TJ electrical. Figure 5 shows the dependance of the payback period on hub height mean windspeed.

Offshore Systems

The shallow water offshore design (figure 1b) is gravity based in 20 m deep water. For corrosion reasons the lower 25 m of tower (above the water line) is concrete. The land foundation is replaced by a larger concrete substructure reaching down to the seabed, a technique suitable for water depths to 50 m. The substructure design is in general accordance with published offshore wind turbine studies (19,20,21). The below waterline mass is estimated at 4100 t before ballasting (at 2 GJ/t for average reinforced concrete).

Construction of the offshore machines requires a number of additions to the land requirement. Some form of dock is needed with cranes and covered areas for subassembly work. The production scenario assumed is batch production of six units at a time, taking one month. After ten years the dock has produced 720 machines (2.88 GW); it is then used for the ten-yearly major overhaul for which the machines are returned to the dock. Provision of the construction base facilities requires 0.60 TJ per machine. The offshore construction requirement is: dock 0.60 TJ, land transport 0.80 TJ, sea transport 2.40 TJ, ballasting 0.10 TJ, sitework (at base) 0.20 TJ, seabed preparation 1.50 TJ. The total is 5.60 TJ (0.05 years), compared with the land machine figure of 7.00 TJ, showing a saving because of the elimination of access roads.

The offshore system requires the use of marine power cable. Figure 6 shows the connection system envisaged. Sub-groups of 25 machines are joined by 6 kV 3-phase cable to a common sub-station. This 100 MW system is connected to land via a 127 MW 132 kV 3-phase cable. Four such adjacent systems are interconnected by a 127 MW cable to provide redundancy. allowing essentially full array power to be transmitted even if one main cable is damaged. (Four single phase cables may be preferable in practice.) Table 10 shows the energy requirement of the 4 MW cable. The installation rate of 288 MW/year requires a total of about 250 km of the two types of cable per year. The manufacturing element is based on the provision and operation of a suitable factory, which we estimate to add 35% to the input material energy requirements. Table 11 (and figure 7) shows the complete offshore grid connection requirement, in which cables outside the array are assumed to be ploughed-in for protection. A 5000 t ship is fully dedicated to cable laying.

Table 12 summarises the offshore system, giving a total of 44.7 TJ (0.37 years). Increase of water depth to 40 m uses an extra 8000 t of reinforced concrete for the substructure, requiring an extra 16 TJ (0.13 years). Beyond 50 m depth a floating platform becomes attractive, for example a tension leg platform (TLP, figure 1c). Table 13 shows the total energy input of a TLP system, in which the platform is estimated to require 10000 t of reinforced concrete (8000 t hull and 2000 t anchors) and 30 t of steel cable for moorings. The total is 59.7 TJ (0.50 years). The use of a steel hull, estimated mass about 1000 t, considerably increases the energy input.

Maintenance

Table 14 shows maintenance energy estimates. We assume that the offshore systems are towed back to the land base for the 10 year major overhaul. Avoiding this would significantly improve the energy gain. Other transport is insignificant. Table 15 summarises the life energy costs. Finally, Table 16 shows the payback periods and complete energy gain (figures 5 and 8).

Conclusions

This detailed process analysis of energy requirements for large wind turbine systims shows that there is a significant contribution from the large number of minor items. Caution should be exercised in comparing these results with those of other systems unless they too are adequately detailed.

For a good UK flat land site with hub height mean windspeed of 7.5 m/s, the initial investment payback period is 0.47 years and the life energy gain is 50.

For a shallow water (20 m) offshore site with hub height mean windspeed of 9.2 m/s, the initial investment payback period is 0.37 years and the life energy gain is 55. The more speculative deep water (100 m) version is rather worse at 0.50 years, with a gain of 39.

Nuclear plant (AGR) has been estimated by Chapman (22) to have a gain of 37. Thus it appears that wind energy systems are extremely competitive from the energy analysis point of view.

References

1 Dixon J C and Lowe R J (1981) Energy analysis of wind energy systems, Open University report in preparation.

2 Roberts F, quoted in Harrison R, Jenkins G and Mortimer N D (1978), Energy analysis in the assessment of the UK wave energy programme, Sunderland Polytechnic EW18.

3 Chapman P F (1975) The energy costs of materials, Energy Policy V3 No1 March 1975.

4 Casper D A, Chapman P F and Mortimer N D (1975) Energy analysis of Report on the census of production 1968, Open University ERG001.

5 Chapman P F (1973a) The energy cost of producing copper and aluminium from primary sources. Open University ERG001.

6 Chapman P F (1973b) The energy cost of producing copper and aluminium from secondary sources. Open University ERG002.

7 Harrison, Jenkins and Mortimer (Ref2) for the electrical generator.

8 Pratt E (CEGB) Private Communication (1981) provided data on transformer design.

9 Charlesworth G (1979) Transport fuels: choice and national implications, PhD thesis, Open University.

10 Varley J S and Harrison R (1977) Energy use in large construction projects,Sunderland Polytechnic EW3.

11 Hemming F (1975) Energy requirements of North Sea oil production, Open University ERG010.

12 Braam J (1980) The relevence of energy analysis in product design, Applied Energy V7 No4 Dec 1980.

13 IFIAS (1974) Energy analysis workshop on methodology and conventions, International Federation of Institutes of Advanced Study (IFIAS) Workshop 6, Stockholm.

14 Poor R H and Hobbs R B (1979) The GE MOD1 WTG Program NASA N80-16457

15 Douglas R R (1979) The Boeing MOD2 WTS rated at 2.5 MW, NASA N80-16457

16 Wind Energy Report, USA, Nov 1980, for data on GE MOD5A.

17 Doman G S (1979) System configuration improvement, NASA N80-16481.

18 Gordon J E (1978) Structures, Pelican (p319).

19 Kilar L A et al (1979) Design study and economic assesment of multi-unit offshore WECS application, ERDA E(49-18)2330.

20 Simpson P B and Lindley D (1980) Offshore siting of WTGs in UK waters. BWEA Second Wind Energy Workshop, Cranfield. Multiscience.

21 Hardell R and Ljungstrom O (1978) Offshore based WTS for Sweden - a system concept study, BHRA Second International Symposium on WECS.

22 Chapman P F (1975) Energy analysis of nuclear power stations, Energy Policy 4 (3) pp285-298 Dec 1975.

TABLES

TABLE 1 BASELINE LAND SYSTEM

		BOEING MOD 2	G.E. MOD 5A	BASE-LINE
Diameter D	m	91.4	106	100
Hub height h	m	61	76.2	80
h/D	-	0.67	0.72	0.8
Rated power P	MW	2.5	4.0	4.0
Rotor loading	W/m^2	381	450	509
Rotor speed	rpm	17.5	12/18	12/18
Design mean(10m)	m/s	6.5	6.5	6.5
windspeed (hub)	m/s	8.8	9.1	9.2
Machine output	TJ/y	35.1	59.4	53.0
	GJ/m^2y	5.35	6.73	6.73
Load factor	-	0.445	0.471	0.42

TABLE 2 BASELINE ARRAY

Machines	-	20 x 5
Spacing (10D)	km	1.0
Installed power	MW	400
Array efficiency	-	0.84
Availability	-	0.95
Transmission eff.	-	0.95
Array peak power	MW	320
Hub height mean speed	m/s	9.2
Array output	TJ/y	4000
Array load factor	-	0.32
Output/machine	TJ/y	40
Offset primary energy	TJ/y	120

TABLE 3 SYSTEM ANALYSIS

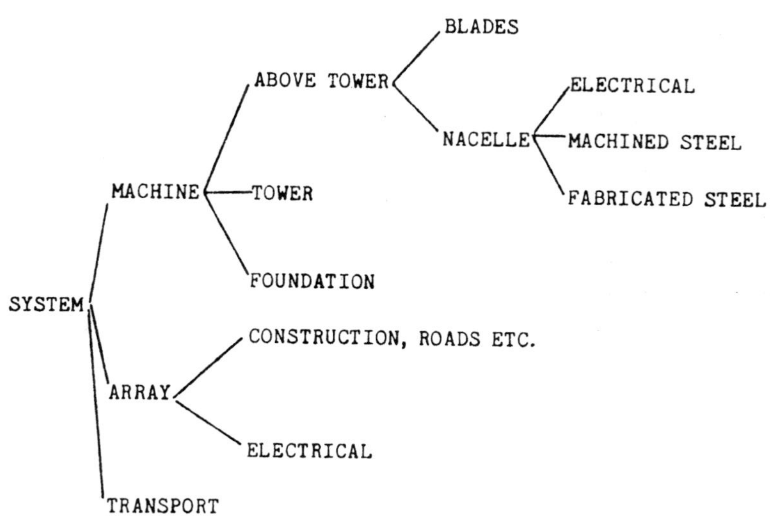

TABLE 4 BASELINE ABOVE TOWER MASS

	kg/kW	tonne	class
Blades (GFRP)	15	60	
Hub (steel)	10	40	machined
Gearbox	7	28	machined
Generator	3	12	electrical
Drive train	3	12	machined
Nacelle structure	9	36	fabricated
Yaw drive (stepping)	2	8	fabricated
Sundries	1	4	fabricated
SUMMARY			
GFRP	15	60	
Machined steel	20	80	
Fabricated steel	12	48	
Electrical	3	12	

TABLE 5 COMPLIANT TOWERS (STEEL SHELL)

		BOEING MOD 2	G.E. MOD 5A	BASELINE
Rated power P	MW	2.5	4.0	4.0
Tower mass	tonne	114	211.4	200.0
	kg/kW	46	52.8	50
Hub height	m	61	76.2	80
Diameter A	m	3.05	3.35	3.5
B	m	6.35	4.57	4.5

TABLE 6 FOUNDATIONS

		BOEING MOD 2	G.E. MOD 5A	BASELINE
Power	MW	2.5	4.0	4.0
Diameter	m	20.4	20.0	20.0
Volume	m^3	441	667 (1)	667
Mass	tonne	1066	1600 (1)	1600
Mass/power	kg/kW	426	400 (1)	400

1) Estimated from indirect data.

TABLE 7 BASELINE MACHINE SUMMARY

		mass tonne	q GJ/t	Q TJ	T year
Above tower (Nacelle 100t Rotor 100t)	GFRP	60	100	6.0	0.05
	Machined steel	80	60	4.8	0.04
	Fabricated steel	48	56	2.7	0.02
	Electrical	12	90	1.1	0.01
Tower	Steel shell	200	50	10.0	0.08
Foundation	Reinforced concrete	1600	2	3.2	0.03
Totals				27.8	0.23

TABLE 8 ALTERNATIVE ROTORS (100m, 4MW)

Material		mass of 2 blades. tonnes	q GJ/t	Q TJ	t years
GFRP (baseline)		60	100	6.0	0.05
Steel		80	56	4.5	0.04
Wood	wood	50	3	2.1	0.02
	epoxy	10	200		
Duralumin		60	110	6.6	0.06
CFRP		18	4000	72.0	0.60
C/GFRP	CFRP	6	4000	28.0	0.23
	GFRP	40	100		

TABLE 9 COMPLETE LAND SYSTEM

	Q/TJ	T/y
Rotor blades	6.0	0.05
Nacelle & hub	8.6	0.07
Tower (comp.)	10.0	0.08
Foundataion	3.2	0.03
Construction (inc. roads)	7.0	0.06.
Array electrical	2.5	0.02
Total	37.3	0.31
Significant variations:		
Rigid tower	+6.0	+0.05
C/GFRP blades	+22.0	+0.18

TABLE 10 MARINE 4 MW 6 kV CABLE XLPE 3-PHASE

Material	Cross section (3 phases) mm^2	Mass kg/m	q MJ/kg	Q MJ/m
Copper	450	4.0	46	184
Polyethylene	1840	2.0	89	178
Steel (armour)	540	4.2	60	252
Material total		10.2		614
Manufacture				215
Total				829

TABLE 11 OFFSHORE GRID CONNECTION (20 km offshore)

	Per 400 MW array	Per 4 MW machine	q	Q TJ
6kV Cable	200 km	2 km	0.83 GJ/m	1.66
6 to 132 kV transformers	4x150 t	6 t	100 GJ/t	0.60
Transformer substations	4x2000 t	80 t	2 GJ/t	0.16
Other (switches etc)				0.20
132 kV Cable	102 km	1 km	3.30 GJ/m	3.30
Cable laying unploughed	222 km	2.2 km	0.08 GJ/m	0.18
ploughed	80 km	0.8 km	0.20 GJ/m	0.16

Total = 6.26 TJ = 0.05 y

TABLE 12 COMPLETE OFFSHORE SYSTEM (SHALLOW, 20 m)

	Q/TJ	T/y
Rotor blades	6.0	0.05
Nacelle & hub	8.6	0.07
Tower (1)	10.0	0.08
Platform	8.2	0.07
Construction	5.5	0.05
Grid connection (20 km)	6.3	0.05
Total	44.7	0.37
Significant variations:		
Depth 40 m	+16.0	+0.13
Offshore 50 km	+4.2	+0.04

TABLE 13 COMPLETE OFFSHORE SYSTEM (DEEP, 100 m)

	Q/TJ	T/y
Rotor blades	6.0	0.05
Nacelle & hub	8.6	0.07
Tower (1)	10.0	0.08
Platform (TLP)	21.8	0.18
Construction	7.0	0.06
Grid connection (20 km)	6.3	0.05
Total	59.7	0.50
Significant variations:		
Steel platform hull	+32.2	+0.27
Depth 200 m	+2.4	+0.02

1) Part concrete tower

TABLE 14 MAINTENANCE (Q/TJ)

		Land	Offshore Gravity (20 m)	Offshore TLP (100 m)
Annual inspection	Parts	0.2	0.2	0.2
5-year mooring change	Parts			1.8
10-year overhaul	Transport	-	4.8	6.0
	Parts	2.0	2.0	2.0
	Paint	0.2	0.2	0.2
	Facilities	0.1	0.5	0.5
	Total	2.3	7.5	8.7

Significant variations:
10 yearly blade replacement 6 TJ per pair

TABLE 15 LIFE ENERGY COSTS (Q/TJ)

	LAND	OFFSHORE GRAVITY (20 m)	OFFSHORE T.L.P. (100 m)
Installed cost	37.3	44.7	59.7
28 inspections (2,3....29)	5.6	5.6	5.6
5 moorings (5,10...25)	-	-	9.0
2 overhauls (10,20)	4.6	15.0	17.4
Total	47.5	65.3	91.7

TABLE 16 SYSTEM OVERALL ENERGY PARAMETERS

		LAND	OFFSHORE (20 km)	OFFSHORE (100 km)
Output electric	TJ/y	40	40	40
Output thermal	TJ/y	120	120	120
First investment energy	Q/TJ	37.3	44.7	59.7
Payback period	years	0.31	0.37	0.50
Total investment energy	Q/TJ	47.5	65.3	91.7
Payback period	years	0.40	0.54	0.76
Life energy output (1)	TJ	3600	3600	3600
Life energy gain	-	76	55	39

(1) Primary energy (thermal) equivalent.

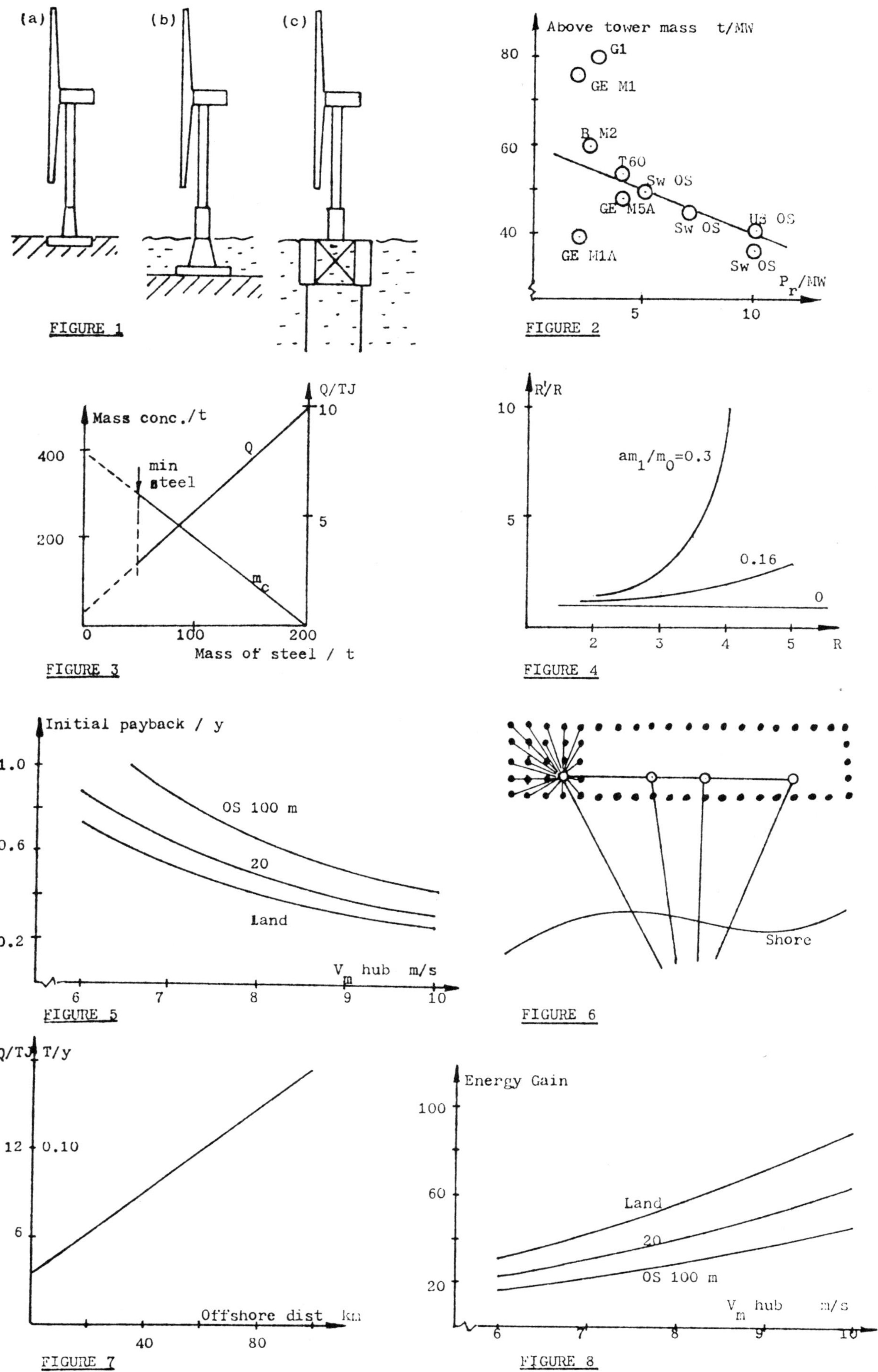
(a)
(b)
(c)
FIGURE 1
Above tower mass t/MW
80
60
40
G1
GE M1
B M2
T60
Sw OS
GE M5A
Sw OS
US OS
Sw OS
GE M1A
P_r/MW
5
10
FIGURE 2
Mass conc./t
Q/TJ
400
200
10
5
min
steel
Q
m_c
0
100
200
Mass of steel / t
FIGURE 3
R'/R
10
5
am_1/m_0=0.3
0.16
0
2
3
4
5
R
FIGURE 4
Initial payback / y
1.0
0.6
0.2
OS 100 m
20
Land
V_m hub m/s
6
7
8
9
10
FIGURE 5
Shore
FIGURE 6
Q/TJ T/y
12
0.10
6
Offshore dist km
40
80
FIGURE 7
Energy Gain
100
60
20
Land
20
OS 100 m
V_m hub m/s
6
7
8
9
10
FIGURE 8

AN APPRECIATION OF THE 10 METRE WIND MATIC AEROGENERATOR OPERATING ON ORKNEY

W.M. Somerville
Chief Development Engineer
NEI Clarke Chapman Engineering

and

W.G. Stevenson
Project Co Ordinator - Wind Energy
North of Scotland Hydro Electric Board

Abstract

This paper describes the 10 metre diameter 22 kW rated Wind Turbine installed at Berriedale Farm on Orkney in December 1980. The horizontal axis turbine has three braced blades of fixed pitch operating upwind of the tower steered by a twin fan tail. Both aerodynamic and mechanical braking systems are included in the protective scheme for the machine and the supply conditions are monitored to ensure a shut down if the network connection is lost. Operating experience has shown that the machine is well suited for unattended operation and that it can produce useful output in wind speeds from five metres per second to forty metres per second. Rated output is achieved in wind speeds above thirteen metres per second. The wide range of operating wind speeds ensures that maximum energy revenue is obtained.

Introduction

The aerogenerator which is the subject of this paper is the smallest machine in a range of four, designed and developed in Denmark in recent years. Other proven machines in the range have rotor diameters of 12 and 14 m equipped with induction generators of 30 kW and 45 or 55 kW respectively. A new machine of 17 m diameter also rated at 55 kW will be produced later this year.

The standard tower consists of three six metre sections giving a hub height of 18.5 metres (60 ft) normally sufficient to lift the turbine clear of the effect of ground features such as hedgerows. When there are no local obstructions to the wind and where strong winds may be expected it is possible to erect the 22 kW machine on a shortened 12m tower made up from the top two sections of the standard tower. This reduces the cost of the tower and the foundations. This option was chosen by N.O.S.H.E.B. for the installation at Berriedale Farm on Orkney taking advantage of an open unobstructed site, see Figure 1.

There are now some seventy machines of this type in operation mainly in Denmark of which approximately 45 are of the 22 kW size.

Description

The Wind Matic aerogenerator has a three bladed, fixed pitch, upwind turbine which is mounted on a substantial hollow steel shaft supported in heavy duty bearings and inclined upwards at an angle of ten degrees. The construction of the turbine and some of the salient features of the machine are clearly shown in figure 2. The blades of the turbine are bolted to the hub and each supported by a steel bracing bar attached to a pylon on the hub.

Additional support is provided by tensioned guy wires stretched between the outer anchor points to prevent torsional vibration of individual blades. The hub itself is a heavy steel fabrication consisting of three folded steel channel sections welded to the boss and supported by two stiffening rings. The pylon is in turn welded to the forward stiffening ring to form the structure shown. Provision is made in the assembly for the control lines which operate the air brakes on the wings. The hub is keyed to the main shaft and further locked by a closing plate which is bolted both to the shaft and hub boss.

The blades of the air turbine are formed in glass reinforced plastic and bonded to a laminated timber beam which extends from the hub to a point at approximately 80% of the blade radius. The timber is coated throughout in glass reinforced plastic to protect it from the elements. The blade is attached by through bolts and clamping pieces which pass through the beam and the shells at the point of attachment of the bracing bar and also where the protected beam lies within the channel section of the hub arm. Moulded within the shell of the cambered rear surface of the blade is a pocket for the air brake assembly mounted near to the tip of the blade. The air brakes can be clearly seen fully extended in Figure 2.

These same control surfaces have an intermediate position where the brake flaps are only partially extended to act as spoilers to limit the power collected by the turbine. The aerofoil section of the wing is a composite of several profiles such as NACA 4412, NACA 4413 and others the tip section pitch angle is approximately three degrees and the wing has a slight warp of nine degrees over its length.

The twin fan tails are a traditional feature of Danish Windmills affording an equal response to a veering or backing wind. These blades, mounted on a common shaft, provide the power to yaw the machine by means of a chain drive and worm reduction gearbox to a pinion which engages with the gear cut rim of the tower top plate.

The nacelle enclosing the gearbox, generator and mechanical control gear is a substantial steel fabrication hot dip galvanised to prevent corrosion. Weather protection is provided by a moulded glass reinforced plastic cover attached by hinges and a folding strut. Four small platforms readily accessible are bolted to the nacelle so that convenient access to all the equipment is available when the cover is raised. There are adequate handholds and attachment points for safety belts. Push buttons are provided in the nacelle to operate the spoiler control and to start and stop the machine for the convenience of the engineer during commissioning and inspection.

Tower and Mounting

The standard tower is of square section tapering from the base to the top plate. The upper two sections are unit fabrications formed from substantial round steel corner members braced with round lattice members all fully welded and hot dip galvanised overall. The lower section, not used in this case, is too wide for road transport and is available as four corner pieces readily bolted together at site.

The nacelle is centred on the top plate by a large hollow pin some 150 mm diameter. The weight is supported by three sliding pads on the underside of the nacelle which ride on the upper face of the tower top plate. The nacelle is prevented from lifting by three keeps which extend beyond and below the rim of the top plate. When mounted the yaw drive pinion engages with the gear cut rim of the tower top plate.

The tower is bolted to a mass concrete foundation block by twelve high strength bolts cast in position whilst accurately located by a template. These bolts provide a facility for setting the tower true and level before grouting the feet.

Nacelle Machinery Arrangement

The machinery arrangement is shown diagramatically in figure 3. The air turbine (1) on its main shaft is supported on two heavy duty double roller bearings (2) and (4) arranged to carry the weight of the air turbine, the aerodynamic thrust and rotational forces imposed in operation together with the weight of the gear transmission (3) and the gear reaction loads. The gearbox, of the shaft mounted type with a step up ratio of 1 to 14.83, is keyed and locked to the main shaft between the main bearings. The gear-box is also supported by a torque reaction arm which allows accurate alignment with the generator. The high speed shaft of the gearbox supports a brake disc (10) and is coupled to the induction generator by means of a flexible rubber coupling (11). Mechanical braking is provided by a lever operated caliper which is arranged to be spring applied (6) and released by a pneumatic cylinder (5) controlled by an electro pneumatic valve. Compressed air is provided by a small single phase compressor (8) mounted on its own receiver located in the nacelle. Compressed air is used to provide power to control the spoilers/air brakes (9). This is accomplished by means of a pair of cylinders (7) in tandem mounted in line with the main shaft. These cylinders are coupled by means of a swivel bearing to a pin which passes axially through the hollow main shaft to a linkage controlling the air brake flaps.

When compressed air is admitted to the cylinders the blade flaps are drawn in until flush with the cambered rear surface of the blade. All three flaps operate together ensuring the braking load is shared. Depressurisation of the spoiler cylinder allows the flaps to rise some 20 mm to break the surface of the blade reducing the "lift" and hence the output. Flow control valves ensure that, when a "stop" is initiated, the air brakes open rapidly without shock to maximum extension to reduce the speed of the turbine before the disc brake takes over to bring the machines to rest. The induction generator (12) is a conventional induction machine wound parallel star the first winding having a relatively high resistance to reduce switching surges on energisation. The main winding is parallelled to the start winding in operation after a one second delay.

Safety Features

In Denmark a subsidy is granted to the purchaser of a windmill of an approved type. To gain such Government approval the windmill must be provided with a number of safety devices intended to insure a satisfactory and safe shut down in the event of a fault and to ensure that the machine can not continue to generate into a disconnected line.

Great attention is paid to the means for preventing a run away and the consequential disastrous failure which would follow. It is expected that both aerodynamic and mechanical braking will be fitted.

The Wind Matic aerogenerator is provided with air brakes and with a disc brake both of which are automatically applied should there be a failure of control power or control air pressure.

If there is a mechanical failure of the power train the rising speed of the main shaft would be detected by a tacho relay initiating shut down. If this relay also fails the rising "lift" on the blades will cause the brake flaps to rise overcoming the control cylinder. Ram effect will fully extend the air brake flaps and operate a trip switch removing control power and causing a "lock out" shutdown.

The supply conditions at the machine are monitored by a series of relays ensuring that the three phase voltages are present, balanced within 5%, and no greater than 110% of the nominal values and also that the frequency is correct $\pm$ 1 Hz. A short time delay ensures that a momentary excursion of the supply caused by local load switching does not shut down the machine. Shut down can be initiated by a manually reset vibration relay fitted in the nacelle, and by a limit switch operated by a mechanical counter if the nacelle turns more than five turns in one direction from the start point. The latter prevents excessive twist in the main cables. There is also an over temperature sensor in the generator winding which will cause a shut down until the machine has cooled to a safe temperature.

In its standard form the machine normally remains off line after a shutdown until it has been inspected. On a remote site much useful generation will be lost if the machine shuts down following a brief supply interruption and the Orkney machine is now being modified to allow auto restart following a supply reconnection subject to all monitored safety circuits being healthy.

Operation

This aerogenerator is aerodynamically self starting. As the speed approaches running speed a speed sensing relay prepares the control relay which compares the generator shaft speed with the supply frequency. As soon as the speed is just above synchronous speed the first power contactor is energised followed after one second by the main contactor. The machine is now generating and the power output will rise with the wind speed. At about 13 m/s wind speed the machine will have reached its rated power and any further increase in power will be sensed by an over current relay which will raise the spoiler flaps to limit the power. Should there be too much power the generator will overheat or the slip speed increase to a point where a shut down trip will be initiated. Normally the machine will stay in production. When the wind speed falls the spoilers retract and the blades fly clean until there is no effective output at which point the generator speed will fall below synchronous speed and be immediately disconnected by the windmill control relay until the wind strength is again sufficient for generation when the cycle will repeat.

Installation and Operation at Berriedale Farm

The foundations were prepared in November 1980 and the mass concrete block poured after suspending the foundation bolts from the levelled base template into the excavation. Provision was made in casting the foundation to carry the power and control cables through the concrete in a duct to a trench and thence to a monitoring "Porta Cabin" sited within the fenced enclosure. During this period the main four core 415 Volt. power cable was laid in a trench from the farm buildings some 300 meters to the North together with a monitoring cable to allow recording of the power consumption on the farm for comparison purposes.

The complete aerogenerator including blades and tower was transported from Newcastle by road-ferry-road direct to site on a single 40 ft. articulated lorry. The parts were unloaded by mobile crane and laid out on the site to facilitate erection on the 15th December.

The erection follows a set procedure which is briefly as follows. The tower top and mid sections are bolted together horizontally and the upper end raised some four feet on a suitable bolster. The nacelle is then lifted by a pair of lugs at the rear and offered to the tower top plate assembling the sliding pads and keeps in the process. The cable twist limit switch mechanism is assembled on the locating tube and the underside of the tower top plate and when the correct adjustment is completed the switch is connected to the junction box within the nacelle. It is convenient at this time to install the main and control flexible cables within the tower guides and in the nacelle. The fan tail blades are then fitted to the hubs and the hubs secured in position at either end of the yaw drive shaft. The three main blades are then bolted to the turbine hub using through bolts and backing plates with lockwashers. The guides for the air brake linkage are also fitted at this time and the linkage made good and adjusted. The blade stay rods are fitted and the stainless steel inter blade bracing wires assembled. The blade supports are then tightened to specified values whilst ensuring that the assembly is true and symmetrical. The whole assembly is then raised to the vertical by crane, lifted and placed on its foundation where it is levelled and secured by the foundation bolts. The control wiring is then made good to the control panel and the power cables installed. Power is then connected and the control functions checked and adjusted if necessary and the machine is ready for service.

In favourable conditions it is possible to erect and commission the machine in one working day, but on Orkney the process took longer as the winds were too strong to permit the final lift until the late evening of the 17th December. Commissioning was delayed by a cable problem until 22nd December when the machine first generated power.

Since this was the first machine of its type to operate in this country the Hydro Board arranged to instal a data logging system at site to record as much significant data as possible on the behaviour of the machine and its control system together with a meteorological log of prevailing conditions. Instruments to record wind speed, wind direction, temperature and barometric pressure were installed locally and impulsing meters to monitor the farm kWhrs, the generator output kWhrs the generator input kWhrs and kVA were connected.

In order to produce the data in manageable quantities the data logging system was arranged to sample wind speed and direction every 10 seconds and "bin" the data against Beaufort numbers 0-15 and sixteen compass points. The store was read onto tape every 30 minutes and recorded along with spot readings of wind speed and direction against a logged time. By this means a daily log reducible to an A4 page format is produced.

It is interesting to observe that, despite the fact that this site is very open and free from obstructions, the wind speed in practically every half hour sample varied over a range of about 2:1 representing a variation in energy flow of 8:1. This variation is quite distinct and separate from the daily variations. To reduce the data to a meaningful figure the distribution of wind speed in each bin was assumed even and the value of the cube root of the mean value of the wind speed cubed found for each half hour sample. The plot of a daily log of wind speed and ½ hour average kW output is shown in figure 4. This data may be replotted as output power versus significant wind speed to give an indication of the machine performance as shown in figure 5 which is self explanatory.

Observing the machine in operation it was seen that as the wind speed approached the rated wind speed the spoilers became active and on occasion were observed to hunt quite rapidly (several times per minute). However, as the wind speed became yet stronger the operation of the spoilers became less frequent and they eventually retracted with the generator producing a substantially constant output at some 25-27 kW. This indicates that the turbine blades are a good compromise for operation in both the "flying" and the "stalled" mode. The machine has stayed "on line" quite happily in some severe winds without shutting down, typically Beaufort 12 on 31st December and the 6th of February.

There have been a number of occasions when generating time has been lost due to storm damage to the feeder and other supply interruptions. The machine requires a visit to restart after such an event and staff were committed elsewhere. The control system is now being revised to allow auto restart after a supply interruption, subject to permission from the safety system which will lock out if manual intervention is required.

To finish with up to date facts; on 2nd March the machine had produced a net output of 13,270 kWhrs during a total generating time of 1,032 hours, an average output of 12.86 kW when on line. During the 71 days since commissioning some 10 days have been lost due to engineering work on the machine and some 7 days due to line fault shut down leaving a balance equivalent to 11 days when there was not enough wind. However, even with low winds some output has been obtained on every day in operation.

More monitoring equipment will be added to the machine to record working loads and stresses later this year and it is expected that much useful experience will be gained.

NEI Clarke Chapman Engineering have been appointed sole concessionaires for the range of Wind Matic machines in the British Isles.

Figure 1. The Aerogenerator on Berriedale Farm, Orkney

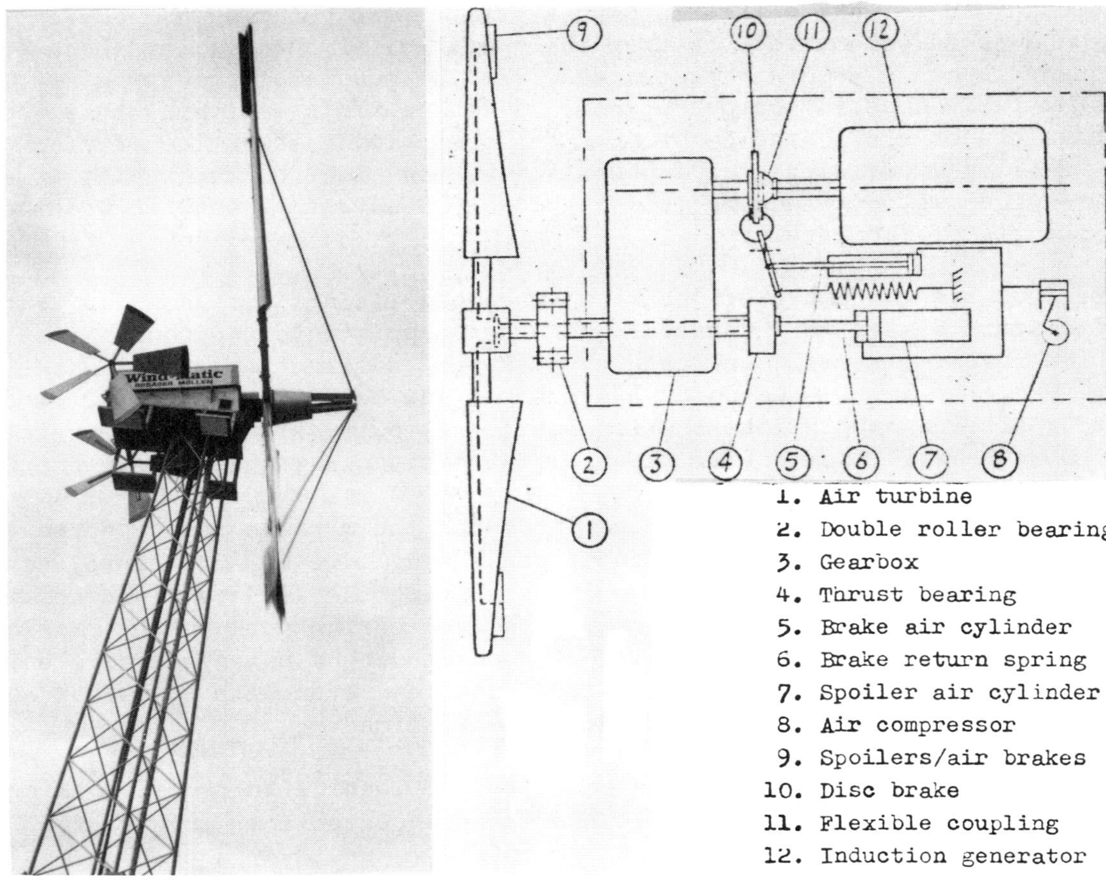

Figure 2. Side View Showing Air Brakes in use.

Figure 3. Diagramatic Machinery Arrangement.

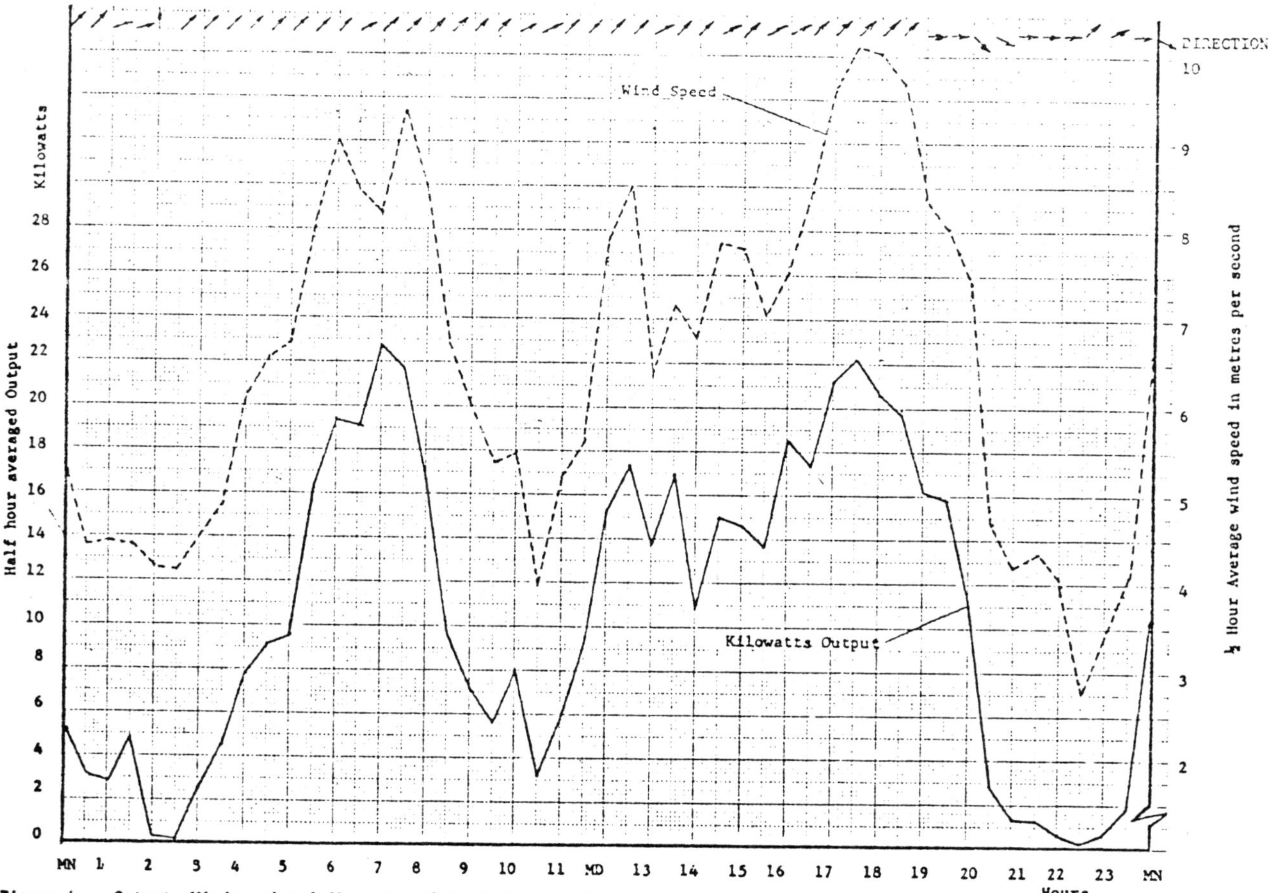

DIRECTION

Wind Speed

Kilowatts Output

Kilowatts

Half hour averaged Output

28 26 24 22 20 18 16 14 12 10 8 6 4 2 0

½ Hour Average wind speed in metres per second

10 9 8 7 6 5 4 3 2

MN 1 2 3 4 5 6 7 8 9 10 11 MD 13 14 15 16 17 18 19 20 21 22 23 MN

Hours

Figure 4. Output, Windspeed and direction plotted from the log for 8th Jan. 81

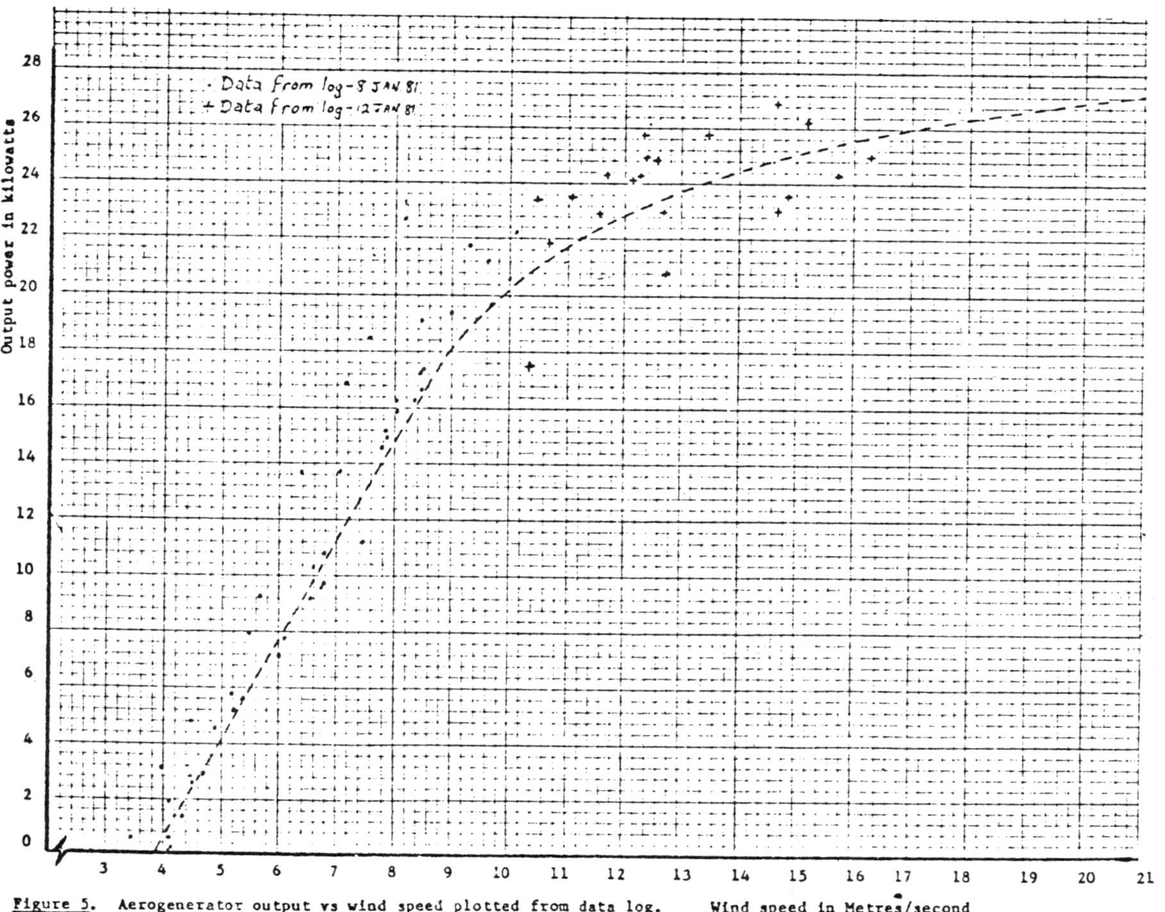

Figure 5. Aerogenerator output vs wind speed plotted from data log. Wind speed in Metres/second

THE RELATIVE ECONOMICS OF WINDPUMPS COMPARED WITH ENGINE-DRIVEN PUMPS

P L Fraenkel
Intermediate Technology Power
9 King Street London WC2E 8HN

Abstract

The Intermediate Technology Development Group Ltd, and its new subsidiary Intermediate Technology Power are developing a modern windpump, designed for low volume economic manufacture in developing countries. It is now in production in Kenya, and is soon to go into production in other countries.

In order to investigate key elements of the design which effect its economy and its competitiveness with diesel pumps (which it is replacing), a computer-based economic model of the windpump has been constructed.

This paper describes some initial results obtained from the model.

Introduction

The Intermediate Technology Development Group (ITDG) initiated a wind-pump development programme towards the end of 1975. The primary objective of this programme was to produce a design suitable for low-volume manufacture in developing countries. Details of the technical design philosophy are given in Fraenkel (1) and of the scope of the project in ibid (2) and (3).

A product range, based on the ITDG design, has been manufactured in Kenya since early 1980 by Bobs Harries Engineering Ltd (BHEL) of Thika, under the trade name "Kijito", see Figure 1. Pilot manufacture has also just been initiated in India by Voltas Ltd., of Bombay, and in Pakistan, by Merin Ltd., of Karachi. Prototypes are under test in Botswana and the Sultanate of Oman and a prototype is also being built in Egypt.

As the design moves into the marketplace, there is an increasing demand for reliable information on its likely physical and economic performance. An investigation and case study on the technology transfer has been completed for the 1981 UN Energy Conference to be held in Nairobi; Adams (4). Our Group also completed an earlier, more generalised study on the cost-effectiveness of water pumping windmills compared with diesel pumps; Gingold (5).

It was decided to take the best information available on the performance of our design and set up a base-line model. Key variables in the model could then be individually changed in order to investigate the sensitivity of the assumptions. Data had also been collected in Kenya and from other sources on the performance and operating costs of small diesel pumping systems, which are a key area of competition for windpumps, and the model was constructed to compare these two types of prime-mover.

The economic model

Cumulatively collected data from Kenya was used as a basis for the model, which correlates daily windpump output against daily windrun. Cumulative data appears to exhibit less scatter than instantaneous data, and to be more universally applicable, since it is the result of averaging energy inputs and outputs over a long period compared with wind gust and windpump time-constants. Windpump output correlates linearly with windrun, since the rotor tip-speed ratio remains within narrow limits and positive displacement pumps produce an output generally in proportion to the number of strokes (or rotor revolutions). The importance of uncertainties in the assumed constants defining the relationship between wind run and water output can be investigated by sensitivity analysis.

The model requires three sets of parameters to be defined, relating respectively to the windpump, to the engine and to financial and physical external effects. These are listed in Table 1. Most of the parameters, and how they fit into the calculation, are self-evident, but a few require explanation. Kenyan cost data, translated into Sterling, have been generally used. Only above-ground costs are considered in the analysis, since below-ground (well) costs are highly site-specific and will be roughly the same regardless of choice of system. It is assumed that water demand is roughly constant, as is the case for human and livestock supplies, (irrigation, where water demand fluctuates considerably demands specialised correlation of crop water demands, rain availability and wind variation through the growing seasons).

The average windrun for the least windy month is used to determine the "firm" output that can be expected. The "Wind Pump Ouput Margin" is an assumed safety margin consisting of the percentage of extra water delivered in an average day of the least windy month of the year if the windrun corresponds precisely with the monthly average. This margin is necessary to allow for errors in assessing the local wind regime and in sizing the windpump as well as for annual and day-by-day fluctuations in the supposed "average" daily wind run of the least windy month, (although a storage tank helps with short-term fluctuations). The choice of margin is at present arbitrary (30% in the baseline model) and can be refined with experience.

The windpump load factor, which the model computes, indicates the ratio of actual production to required production. No value is assigned to the surplus that is pumped during windier months.

Discount and inflation rates are held constant over the 20 year period assessed by the model and capital costs are annualised as equal payments in the usual way. The amortization period used to annualise the capital costs is the number of years the machine will run, calculated on the basis of specified operational life (in hours running time) and the actual running time per year required to satisfy the specified demand.

The model then computes various results, also indicated in Table 1 and produces a table of annual total running costs (in 1981 £Sterling) for the windpump and the diesel system for every year from 1981 to 2000, broken down into annualised capital cost, maintenance cost, and for the engine, fuel cost, plus the cost ratio.

Finally, in order to obtain an indication of how the lifetime costs of either type of system option compare, the Net Present Value (NPV) of all the cash flows is calculated, again broken down for each system into capital, maintenance and fuel NPVs and the ratio of total system NPVs is determined. This latter result in effect compares the lifetime cashflow values, discounted to the present day, of the two options.

The model was operated by changing selected variables from those used in the base-line model by various percentages in order to investigate the system response.

Discussion of results obtained

Some selected results of the annual running costs from 1981 to 2000 are presented in Fig 2. The solid lines represent windpump systems while the broken lines are diesel engine pumps. The fuel needs of diesel power cause a marked exponential rise in costs, while the low running costs of the windpump result in large step changes if and when a new unit is purchased at an inflated price.

Reasonably modest initial fuel costs are assumed of 20p/litre (90p/Imperial Gallon) and the inflation rate for fuel for the base-line model is taken as 20% (compared with an actual rate of about 25% for the last decade). Run 2b in Figure 2 shows that even if the fuel inflation rate is taken as being only 15%, annual running costs for the diesel still increase by a factor of 10 in 20 years, while a continuing rate of 25% (run 2a) would result in astronomic running costs.

The same figure also shows the sensitivity of windpump operating costs to windspeed and to discount rate. If the windspeed in the least windy month averages 30% less than expected or hoped for, the costs to produce sufficient water by wind power are approximately doubled (either by the use of a larger machine, which may in fact yield some economies of scale to mitigate this effect) or by using two windpumps on two wells.

Figs.3 and 4 are sensitivity curves showing the effects of varying the parameters indicated on present day unit output costs and on NPV ratio. It is clear that windspeed is a critical factor (the ratio of mean wind speed for the least windy month to annual mean wind speed was held constant at the ratio of 2:3 when wind speed parameters were compared). This shows how important an adequate knowledge of the wind regime is for sizing windpumps; when in doubt a larger than necessary machine must be used or standby power of an alternative kind must be made available. However, the baseline windspeeds, representing a least windy month average of 200km/day (or 2.3m/s or 5.2 miles/hr) are reasonably widely to be found. According to Figure 4, windpower would retain lower lifetime costs unless the wind speed fell over about 35% below the base-line (to about 1.5m/s or 3.4 miles/hr). This seems to confirm a generally applied rule based on experience that windpumps are viable providing the local annual mean windspeed is over about 8km/hr (5 ml/hr).

Windpump operational life is also a critical factor; a 50% reduction in life causes a 75% increase in unit costs. The model is perhaps a little pessimistic about the windpump in writing it off completely once its notional life is over, since unlike with an engine, it would normally pay simply to replace the windpump mechanical components and retain the tower and possibly the rotor and tail for another "life"; this is certainly the practice in Australia, where many windpumps have been in daily use for many decades. The high sensitivity to operational life assumptions points to the value of producing a high quality windpump and suggests that any supposedly "cheap" machine will in the end be expensive unless it can function for over 30000 hours without major replacements.

For the diesel pump, fuel price inflation assumptions are critical (see Fig.4). The inflation rate would have to be in low single figures before the NPV of the engine would be less than that of the base-line windpump.

The "minimum output safety margin" value does not too radically effect windpump economics. If it were felt that 30% is too small a margin for safety, then increasing it by half to 45% would yield a notional unit output cost increase of 25%. In practice, a larger safety margin would require a larger windpump at a given site, and this would attract economies of scale which are not addressed by this model - so the resulting cost increase would if anything be lower.

Maintenance costs are such a small proportion of total costs for the windpump that they could be increased radically with only marginal effects on the total costs; there is consequently little to be gained financially by seeking totally maintenance-free operation (or very limited maintenance operation) other than convenience for the owner. Maintenance costs are only slightly more significant with the diesel pump.

Conclusion

Important effects that are neglected at this stage are the shadow costs and benefits incurred by a locally manufactured product using a very small foreign currency contribution in its lifetime cash flow compared with the imported engine powered pump which burns many times its own value in imported fuel, all to be paid for with hard currency, (the NPV of the baseline engine fuel costs up to the year 2000 is no less than 72% of its total NPV!).

One of the main reasons stated by purchasers of Kijito windpumps for having decided to buy a windpump rather than another engine-pump is a serious worry about whether they can count on diesel fuel being permanently available in future, at any price. Therefore, whether or not windpumps are cheaper, the wind is seen as a more secure energy resource than oil in many places. However, the indications so far obtained, are that the windpump, under Kenyan conditions, does have a significant economic advantage in the longer term over engine-driven pumps, and in many cases it appears to have an immediate economic advantage.

Acknowledgements

Funding for the development of the windpump this paper refers to was provided by Christian Aid plus the resources of our various overseas collaborators in the project. The Overseas Development Administration and Intermediate Technology Industrial Services are funding and managing the field testing programme in Kenya mentioned in this paper. Without this help this work would not have been possible.

References

1. Fraenkel, P.L.,"An international development programme to produce a wind-powered water-pumping system suitable for small-scale economic manufacture", Proc. 2nd Int.Symposium on Wind Energy Systems, Amsterdam, 1978. BHRA Cranfield 1978.

2. ibid "The Intermediate Technology Windpump Development Programme", IEE Joint Professional Group S1 and BWEA Colloquium on Windpower, I.E.E., London, November 1979.

3. ibid "The ITDG international windpump programme; engineering design considerations used in developing a windpump system for small-scale manufacture and use in underdeveloped arid or semi-arid regions", Proc. Indian Academy of Sciences, Vol.C2, Pt.1, March 1979.

4. Adams, W.E., "The ITDG International Windpump Programme: a Case Study of Technology Transfer", (private paper commissioned by the UN (New York) from IT Consultants Ltd, London, December 1980.

5. Gingold, P.R., "The Cost-effectiveness of water-pumping windmills", Intermediate Technology Industrial Services, Rugby, 1979.

TABLE 1 THE VARIOUS PARAMETERS USED IN THE ECONOMIC MODEL SHOWING THE BASELINE VALUES THAT WERE ASSUMED AND BASELINE COMPUTED RESULTS

WIND PUMP PARAMETERS		ENGINE PARAMETERS	
Wind rotor diameter	7.3m	Rated power	6 kW
Output margin (least windy mth)	30%	Derating factor	.75
Capital cost (yr 1)	£ 5250	Unit capital cost	£ 280/kW
Maintenance cost (yr 1)	£ 52	Maint.cost / hr run	30p
Est.operating life	50000hr	Est.operating life	5000hr
Subsystem (pump) effic.	60%	Subsystem efficiency	50%
		Engine efficiency	25%

EXTERNAL PARAMETERS		COMPUTED RESULTS	
Discount rate	10%	Min.firm water output	24.4cu.m./day
General inflation rate	10%	Annual energy demand	1216 kWh
Fuel price inflation rate	20%	Windpump system life	16 years
Av.wind-run least windy month	200km/day	Av.windpump output least windy mth	34.93 cu.m./day
Av.wind-run for whole year	300km/day	Av.windpump output for whole year	63.94 cu.m./day
Water delivery head	50m	Windpump load factor	.38
Fuel cost at year 1	20p/litre	Engine capital cost	£ 1680
		Engine fuel consumption	1945 litre/year
		Engine running time	540 hr/year
		Engine operational life	9 years
		Capital cost ratio in year 1	3.1
		Cost of water in year 1: wind pump	8p/cu.m.
		Cost of water in year 1: diesel pump	9p/cu.m.

Figure 1:

Kenyan manufactured "Kijito" windpump, with 7.3m diameter rotor on a borehole near Nanuki.

The analysis in this paper relates to a machine of this type and size.

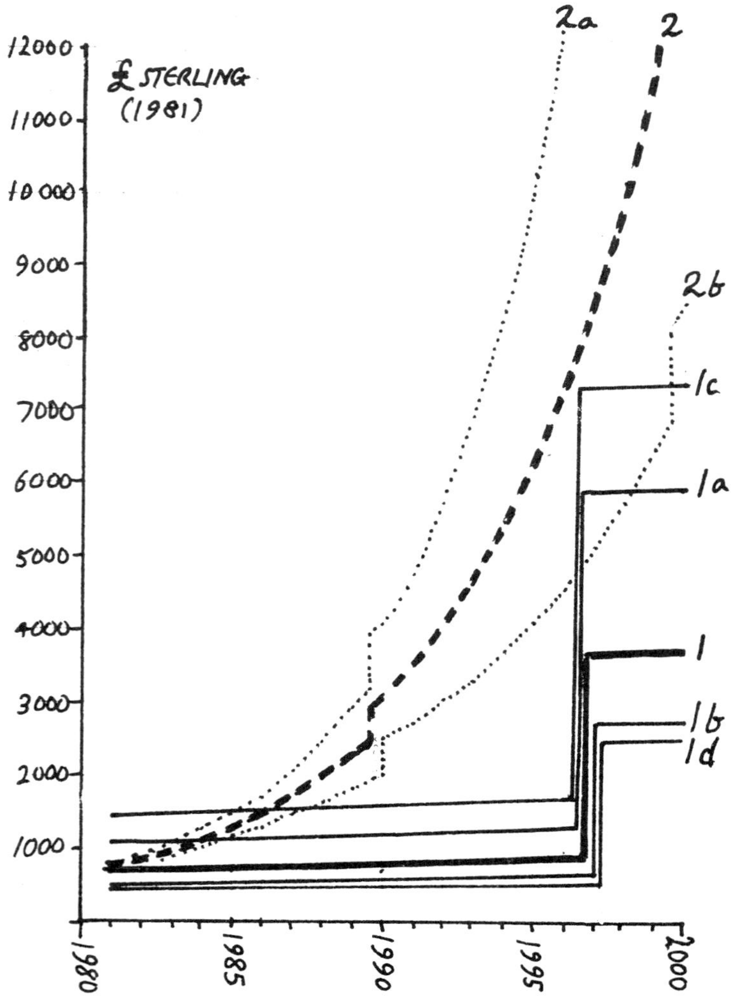

£ STERLING (1981)

12000
11000
10 000
9000
8000
7000
6000
5000
4000
3000
2000
1000

1980 1985 1990 1995 2000

2a 2 2b 1c 1a 1 1b 1d

Figure 2:

Annual total operating costs in 1981 value pounds Sterling for the period 1981 to 2000 for a 7.3m rotor "Kijito" windpump and a 6kW diesel pump.

Legend:

1 baseline windpump
1a as 1 with discount rate doubled
1b as 1 with discount rate halved
1c as 1 with windspeeds reduced by 30%
1d as 1 with windspeeds increased by 30%

2 baseline diesel pump
2a as 2 but fuel price inflation up to 25% pa
2b as 2 but fuel price inflation down to 15%pa

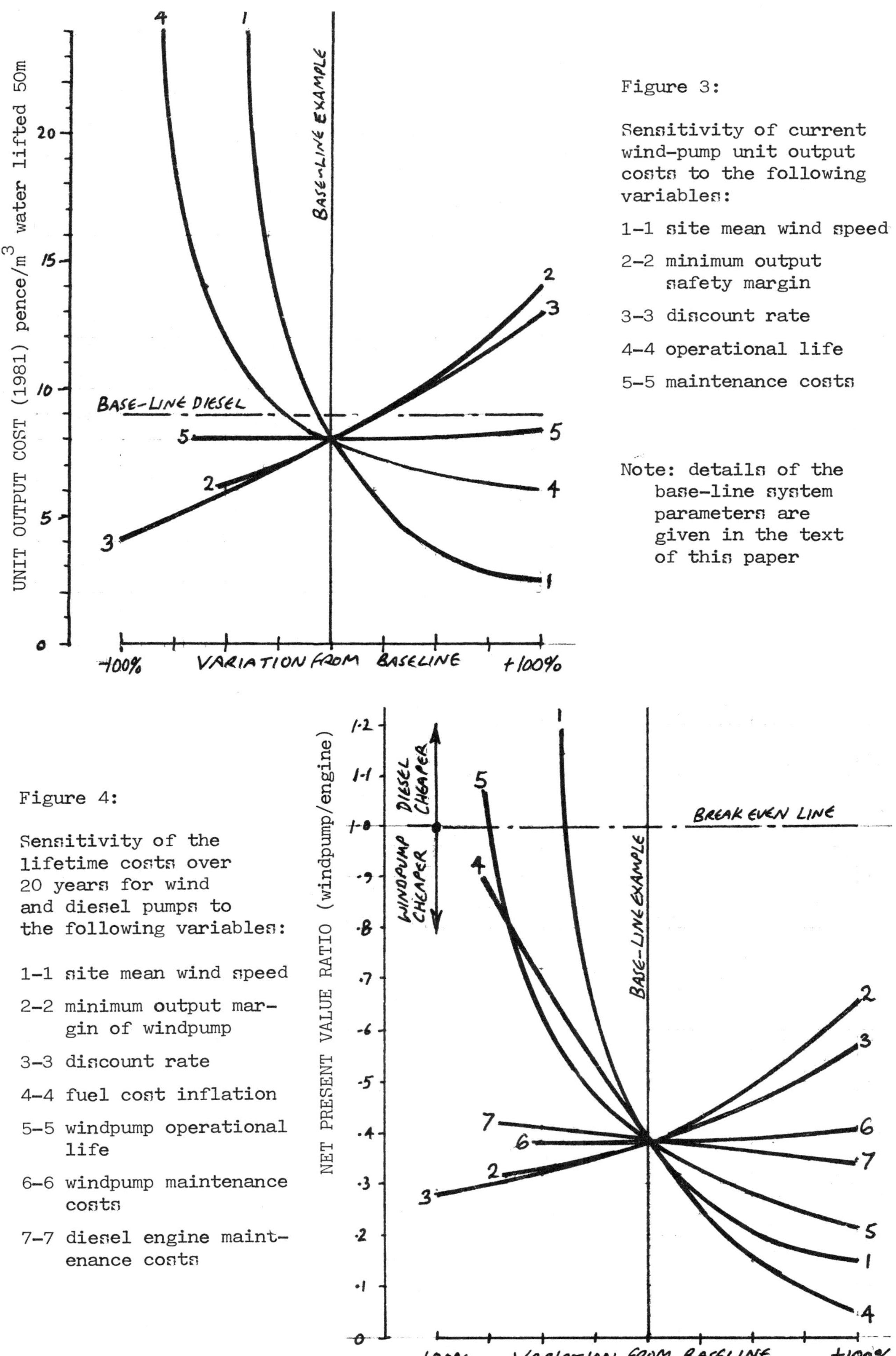

Figure 3:

Sensitivity of current wind-pump unit output costs to the following variables:

1-1 site mean wind speed

2-2 minimum output safety margin

3-3 discount rate

4-4 operational life

5-5 maintenance costs

Note: details of the base-line system parameters are given in the text of this paper

Figure 4:

Sensitivity of the lifetime costs over 20 years for wind and diesel pumps to the following variables:

1-1 site mean wind speed

2-2 minimum output margin of windpump

3-3 discount rate

4-4 fuel cost inflation

5-5 windpump operational life

6-6 windpump maintenance costs

7-7 diesel engine maintenance costs

WIND TURBINE RESEARCH AT NAPIER COLLEGE

W S Bannister and S Gair
Napier College, Edinburgh

Abstract

A straight bladed vertical axis wind turbine is now on a test site and a test programme is now in hand to investigate the characteristics of various turbine configurations. The main line of investigation will be to examine a low aspect ratio, high solidity configuration. Composite glass reinforced fibreglass blades, of NACA 0015 section, are being used initially. The instrumentation includes a microprocessor system which will enable instantaneous measurement of all test parameters to be made and recorded. This system, when installed will enable turbine characteristics to be obtained relatively quickly and provide data on blade and cross-arm stressing. An outline of operation of the wind turbine from a desktop computer is given together with two methods of maximising the energy yield when mechanical to electrical energy conversion is achieved via a permanent magnet type generator. An induction generator scheme is reviewed whereby an isolated induction machine receives its magnetisation current from a synchronous compensator and generates fixed voltage, fixed frequency from a varying wind source to supply a varying load demand.

Introduction

At Napier College work was started on a vertical-axis wind turbine in 1978. Possible alternatives to the basic Musgrove design were presented to the BWEA in 1979. Similar ideas, for example, of increasing the solidity to improve performance were reported by Musgrove and Mays (1) and also the effects of low aspect ratio were arousing attention. In order to investigate the performance of a vertical-axis straight bladed wind turbine over a range of solidity and also with blades of various aspect ratios, it was decided to set up a test facility with appropriate instrumentation. The wind turbine, already constructed at Napier College, would form the basic test installation thereafter the equipment would be developed as the need arose. In parallel with the work on the basic turbine design, instrumentation is being developed to measure the parameters associated with the turbine performance. These parameters are monitored by a micro-processor with a very fast programme turn round time. Data sampling speed can be pre-fixed to satisfy test requirements and the information so gathered stored on magnetic tape. The alternatives of producing electrical power from the wind turbine are being looked at from the point of view of achieving optimisation of two distinct load categories, (i) space heating load, (ii) mains voltage and frequency type loads.

Wind Turbine

The turbine shown in Fig.1, is supported on a steel tube which is fitted to a heavy base with a tilting mechanism to enable easy access for modifications to the blades or cross-arms. A central spindle 25 mm diameter is used to transmit the torque from the turbine to the base of the vertical tube. The height of the turbine is approximately 4 m from ground level to the bottom of the blades.

The cross-arms are aerofoil shaped and approximately the same chord length as the blades. The fixing of the cross-arms to the central spindle is such that cross-arms of different length, that is turbine radius, can easily be fitted.

The blades presently available are of foam filled fibre glass contruction with a NACA 0015 section. The chord is 0.25 m and the maximum length is 2 m giving a maximum aspect ratio of 8. For lower aspect ratios the blades will simply be cut to size. It is hoped to obtain extruded aluminium alloy blades at a later date which will extend the range of the tests.

The electrical power take off is available at ground level and the generator will be fitted after initial tests using the Sharpe (2) technique, i.e. the turbine being loaded purely by its own inertia and being allowed to accelerate through a permissible speed range.

Test Instrumentation

The parameters such as wind velocity, wind direction, angular velocity and angular acceleration of the turbine rotor, strain, electrical power output, etc, are to be monitored using voltage transducers. These voltages will be fed into a sixteen channel multiplexor, each channel in turn will be read and passed into a memory location in a microprocessor via an analogue to digital convertor. When the operation is complete, a clocking pulse will move the multiplexor to the next channel. As the channels are read the data will be stored. The operation is then repeated and the new results stored in other locations. The sampling rate will be programmed in the microprocessor to satisfy the particular test requirements. On completion of an experimental test, information will be transferred to magnetic tape for future analysis.

The hardware consists of a basic microprocessor (M6800 D2 kit) operating from an 8-bit analogue input system with a combined analogue to digital convertor (MP 21 Burr-Brown). A simplified block diagram is shown in Fig. 2.

Speed of operation and flexibility are the main advantages of using this dedicated system. The expected turn round time, that is for sixteen readings, is about 20 milliseconds, but this will obviously depend on the programme time.

The MP 21 is compatible with the M 6800 and very short programmes will be required, around 60 hexadecimal instructions. For fast changing parameters, more than one channel may be used and all sixteen channels may be used if it is required to study any one parameter in detail. Delays of 30 microseconds to several hours may be introduced between each set of readings and the programme may be used to accommodate various run times. Initially 16 k of storage will be available, this enables sixteen sets of a thousand results per set of data to be stored for any one test run, the minimum time for a run being 20 seconds.

This equipment is now being assembled and will be tested in the laboratory before being used at the wind turbine test site.

Electrical Power Take-off

Considerations for electric power generation are being looked at from the point of view of satisfying two separate load requirements.

(a) the need to operate the test turbine on site near maximum power coefficient (C_{pmax}) in an electrically controlled manner such as to maximise the energy yield from the turbine.

(b) the feasibility of generating electrical power at regulated voltage/ regulated frequency from an isolated induction generator (i.e. no local grid connection available).

1. The choice of generator and electrical control required has been based on the need to convert, as efficiently as possible, the power output of the wind turbine and simultaneously control the load in such a manner as to always present the wind turbine with a power α (rotational speed)^3 loading characteristic (i.e. $P \alpha N^3$). The final load demand being stipulated as nothing more sophisticated than a space heat requirement in which the actual voltage and frequency output of the generator is of no consequence. To achieve this aim, a permanent magnet type generator was selected. In the absence of field control for the generator the two remaining control possibilities are now under consideration, these being, (i) load control, and (ii) electrical transmission control.

   (i) Load Control - the power V rotational speed capability of the generator when supplying discrete resistance loads is shown in Fig. 3. A $P \alpha N^3$ characteristic representing the predicted output of the wind turbine can be super-imposed upon the electrical generator characteristics and positioned to give the optimum step wise loading (i.e. $P \alpha N^3$) of the wind turbine. The final positioning of the wind turbine characteristic will determine the gearbox step-up ratio required. Operation within the maximum current loading of the generator can be set and the discrete resistance loading steps determined so that for a particular wind speed-band, operation of the turbine does not move too far away from C_{pmax} nor is the turbine ever stalled by the electrical load. Electrical sensing of the wind speed is carried out at the generator terminals by measuring the output frequency of the generator (this being directly proportional to shaft speed. i.e. frequency = shaft speed x pole pairs). This signal when converted to a voltage level by a frequency-to-voltage converter is used to drive a level detector which in turn performs the load switching function. By suitable adjustment 'dead bands' are created in order to minimise 'hunting' of the stepped resistance load due to gusts in the wind conditions, commensurate with acceptable deviation from C_{pmax}. This method of control has the merits of being cheap, simple and effective for this particular type of electrical load.

   (ii) Transmission Control - a more effective technical solution to the problem of obtaining a wind turbine load match is to use a controlled rectifier in the output of the generator, as shown in Fig. 4. Here by suitable variation of the firing (control) angle (α) of the thyristors in the rectifier unit, it is possible to follow a $P \alpha N^3$ law provided signal measurement and speed of response can be achieved within the time taken for the wind turbine to react to transitory wind changes (the time constant of the system is proportional to the inertia of the rotating mass, the electrical time constant of the generator, the load and control elements and inversely proportional to the wind speed (3)). Although maximising the energy yield in terms of a matched output this system of control has the drawbacks of being both expensive (since the controller is now situated in the power line) and technically complicated.

However, notwithstanding such drawbacks, in an attempt to achieve our objective of maximising the energy yield from the wind turbine it was decided to go ahead with such a system using a controlled rectifier, the control angle of which could be varied by means of a stepping motor operated from a programmable drive unit with IEEE interface BUS, the stepping unit (the listener) being operated from a PET computer (the talker). When control of the device is complete the computer closes down that device and may move onto some other device connected to the BUS. In this way several devices in an automated system may be controlled from a single computer. The computer stores the programme for operating the stepping motor and increments or decrements the stepping motor according to signals of wind speed and shaft speed such as to maintain constant tip speed ratio conditions for the wind turbine.

2. Induction Generator Synchronous Compensator System - problems of operating an induction generator remote from a grid supply point have been investigated in recent publications (4,5) and a solution to the problem of providing magnetisation current for the generator from static capacitors is proposed, subsequent control of voltage and frequency being achieved through rectification and inversion in the power line or simply rectification to produce a d.c. output. An alternative scheme employing an induction generator/synchronous condenser system has been proposed (6) and it is felt that this system offers interesting possibilities for remote locations requiring mains voltage and frequency where synchronous generators/diesel turbine units are already in operation. The layout of such a scheme is shown in Fig. 5. Start-up of the system would be as follows:-

   (i) start diesel engine,

   (ii) increase field current of synchronous generator and thereby start induction machine as a drive motor (useful if wind turbine is a vertical axis machine and not automatically self starting).

   (iii) wind turbine operates as an aerodynamic machine and raises the shaft speed of the induction machine so that it enters the generating mode.

   (iv) de-clutch and shut down diesel engine.

   (v) the synchronous compensator supplies reference magnetisation current to the induction generator which in turn, in addition to supplying the load, delivers real power to the synchronous machine, to replenish windage and friction losses and maintain its rotation at synchronous speed.

Control of the system voltage and frequency is brought about by variation of the synchronous compensator field current (I_f) and some form of rotor power control in the slip-ring induction generator. The system is thereby capable of sustaining regulated voltage, regulated frequency from a varying wind source, in addition to a varying electrical power demand and makes effective use of (a) a lightly loaded existing synchronous machine, and (b) a robust induction generator which can operate over a wide range depending upon the control method employed.

In the electric drive industry, d.c. drives and inverter control drives for induction machines are inherently wide speed range and are rated to carry the maximum drive power no matter the speed range required. Their cost thus increases with speed range and the requirement that they be connected into the main power line as a controller would be incompatible with the use of a synchronous compensator providing reference magnetisation current to the induction generator as in this instance. However, in speed control systems which vary the rotor power extraction, the rating, and hence the cost, of this equipment does reduce in proportion to the speed range. The Kramer and Scherbius machine drives and the Schrage motor are based on this proncipple, but are all "two machines in one", all have commutators, and all have losses associated with the extra brushgear and windings, reducing their efficiency.

A more modern configuration, based on d.c. thyristor drive equipment, is the 'static' Kramer drive shown in Fig. 6. The rotor slip power of a standard type of 3-phase wound rotor induction motor is recovered and fed back to the mains. The 'static' in the title refers to the fact that all recovery equipment is static (thyristors, diodes, transformers, etc). The economics of such a drive have made it attractive for large drives requiring only a short speed range (for reasons of low power factor, down to 60% of rated speed) and achieving overall efficiency of 95%, hence its consideration as a controller in this instance for use with an induction generator.

Operation - The rotor voltage characteristic of an induction machine varies with speed (slip S) as shown in Fig. 7. Similarly, if the rotor a.c. voltage (V_r) is full-wave rectified the resulting d.c. voltage (V_{dc}) is also proportional to slip. If a source of d.c. voltage such as a battery, the armature of a d.c. motor, or a suitable thyristor inverter is connected to the d.c. output of the rotor rectifier, then the induction machine must run at a reduced speed in order to develop sufficient voltage to overcome that impressed at the d.c. terminals to pass rotor current, and hence produce torque. It has been shown (7) that variation of the inverter control angle (α) in the range $90^\circ \leqslant \alpha \leqslant 160^\circ$ will vary the speed of the induction machine over a range 50-100% of synchronous speed.

The Kramer system of control does not give the same torque/speed curves as the more commonly adopted method of insertion of rotor resistance. To emphasise this point Fig. 8 shows torque/speed curves for various values of inserted rotor resistance, and also for various values of impressed d.c. link voltage. The difference is plain. In the static Kramer system the value of d.c. voltage is adjusted by the control loop to bring about the required system conditions of voltage and frequency.

A series of laboratory tests have been carried out with the machine set operating into an isolated grid to ascertain the variation of system voltage and frequency as the load demand changes under the operating constraints of synchronous compensator field current control and wind turbine shaft speed control. These results are shown in Fig. 9, from which a family of curves with ordinates of field current and induction generator shaft speed have been extrapolated, Fig. 10, showing their effect on frequency and line voltage. The results show that the frequency and system voltage are each dependant on synchronous compensator field current and induction generator speed.

Preliminary test work is currently in hand utilising a mercury arc rectifier (as a temporary substitute for a static inverter) to test the system operation with variable delay angle in the inverter bridge. The tests are aimed at determining the range of wind turbine speed which can be accommodated by the induction generator controller (α) in order to maintain a fixed voltage, fixed frequency on the grid.

Conclusions

Investigation of operation of the wind turbine from a small computer is worthy of consideration in view of the flexibility of control which it can offer for test purposes. There is a need to evaluate the technical merits of solid state transmission line controllers as opposed to simpler, cheaper on-off load controllers from the point of view of increased energy capture from the wind turbine.

The advantages of electrical energy conversion using an induction generator are many. The solution to the problem of operating an induction generator remote from a supply grid lies with the choice of using either static capacitors for self excitation with solid state controllers in the main power line or the use of a synchronous compensator to provide reference magnetisation with control carried out on the machine secondaries. The technical and economic aspects of both systems require further evaluation.

Acknowledgements

The authors gratefully acknowledge the many contributions made by colleagues on the Wind Energy Group at Napier College and to other workers in the field of wind energy research. Thanks are also due to the Science Research Council for their support via grant Ref. GR/B52717. The work is being carried out at Napier College, Edinburgh and is being supported by Lothian Regional Council.

References

1. MUSGROVE, P J and MAYS I D. "Development of the variable geometry vertical axis windmill". Second International Symposium on Wind Energy Systems. Paper E4 organised by Brit. Hydromech. Res. Assoc., Amsterdam October 1978.

2. SHARPE, D J. "A Theoretical and Experimental Study of the Darrieus Vertical Axis Wind Turbine", Kingston Polytechnic Research Report, Kingston upon Thames, October 1977.

3. FRERIS, L L. Discussion on "Permanent magnet a.c. generators" and "Operation of self-excited generators for windmill application", Proc. IEE, 1980, 127, Pt. B, No. 4.

4. ARILLAGA, J and WATSON D B., 'Static power conversion from self-excited induction generators', Proc. IEE. 1978, 125, (8), pp743-746.

5. WATSON, D B., ARILLAGA, J and DENSEM B E., 'Controllable d.c. power supply from wind-driven self-excited induction machines'. Proc. IEE. 1979, 126, (12), pp.1245-1248.

6. OOI, B T and DAVID R A., 'Induction generator/synchronous condenser system for wind turbine power', Proc. IEE. 1979, 126 (1), pp 69-74.

7. MURPHY, J M D., 'Thyristor control of a.c. motors', Pergamon Press, Chapter 10.

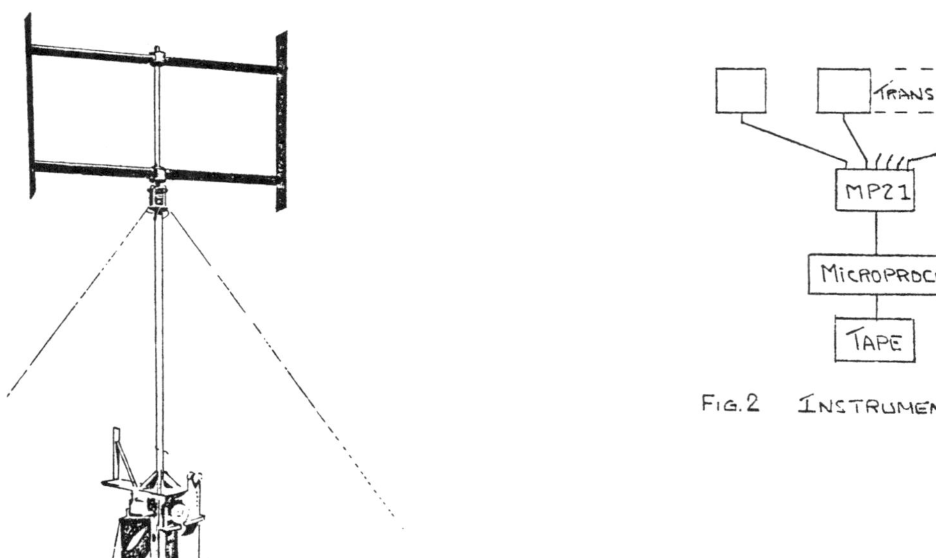

FIG.2 INSTRUMENTATION

FIG 1 WINDTURBINE

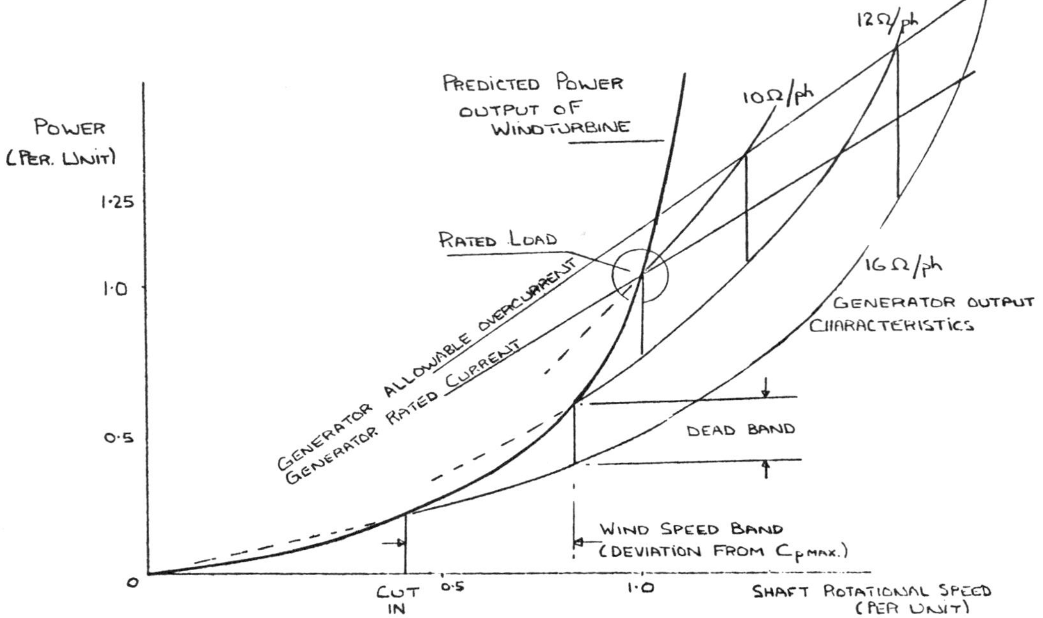

FIG.3 WINDTURBINE/GENERATOR OPERATING CURVE

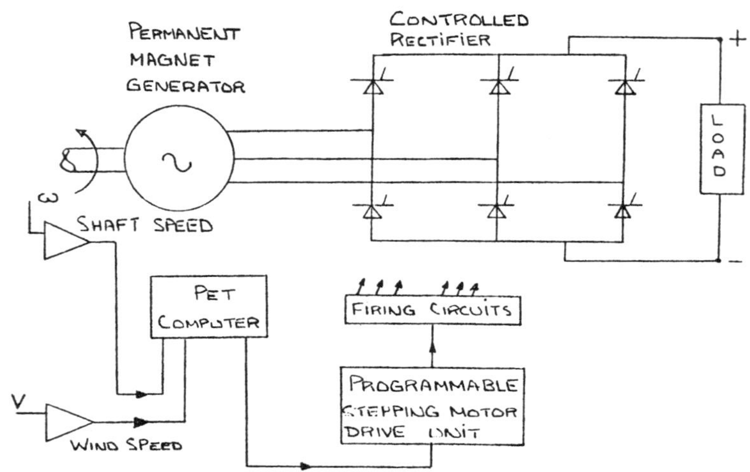

FIG. 4 - POWER TRANSMISSION CONTROL

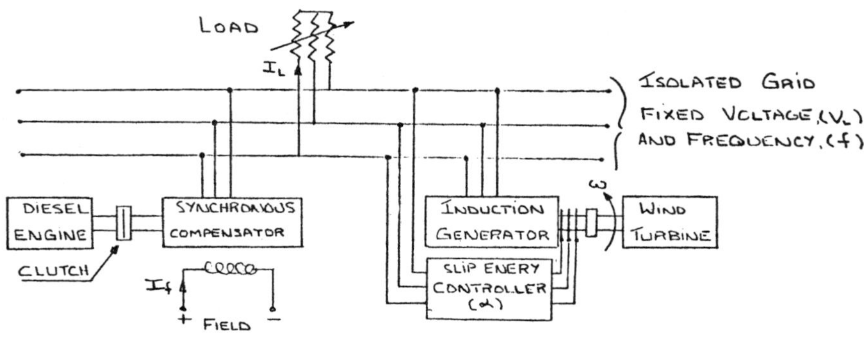

FIG.5 – INDUCTION GENERATOR SYNCHRONOUS COMPENSATOR SYSTEM

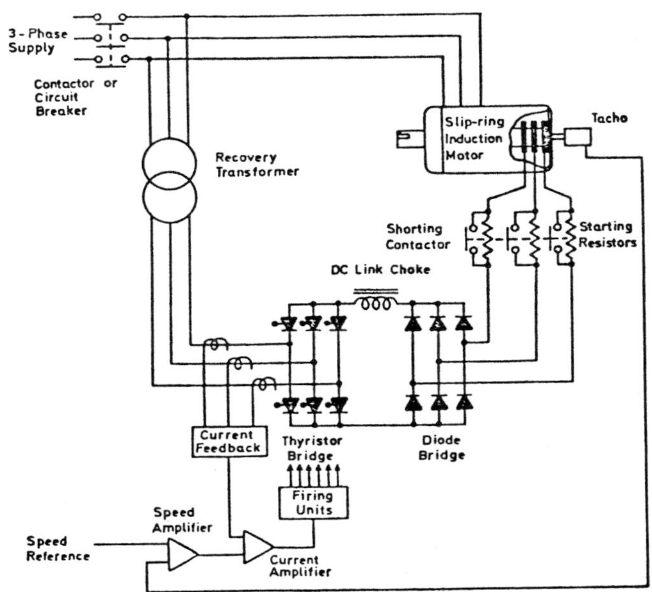

FIG. 6 Basic Schematic of Static Kramer Drive

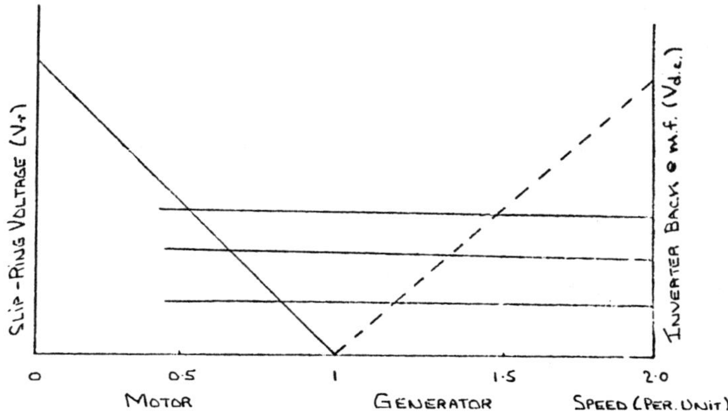

FIG.7 ROTOR VOLTAGE CHARACTERISTIC

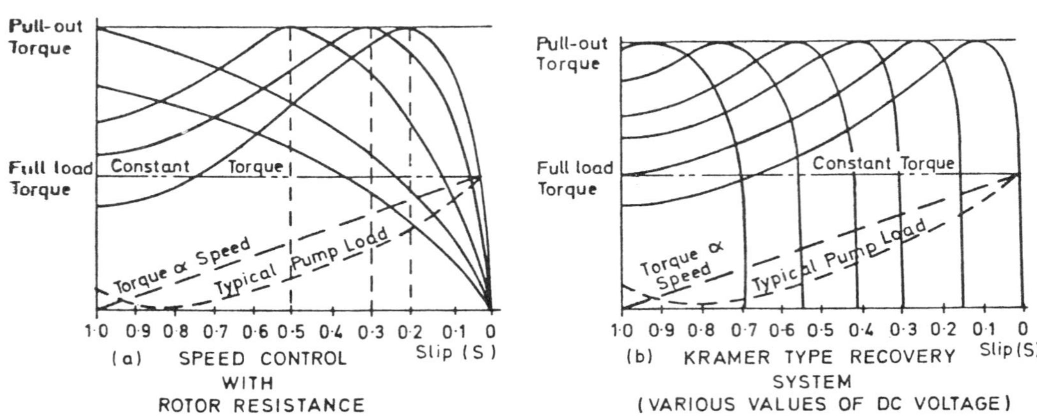

Fig. 8 Torque/Slip Characteristics

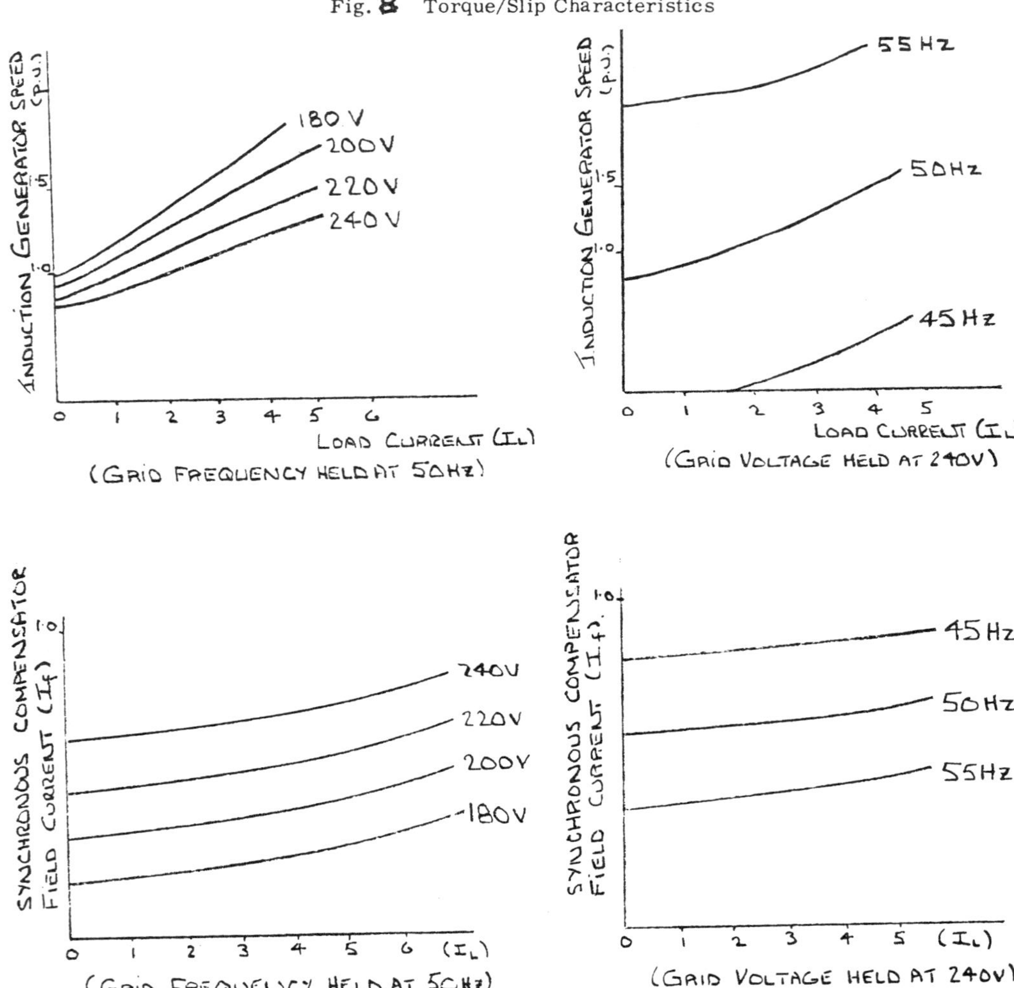

FIG. 9 VARIATION OF VOLTAGE + FREQUENCY WITH LOAD DEMAND.

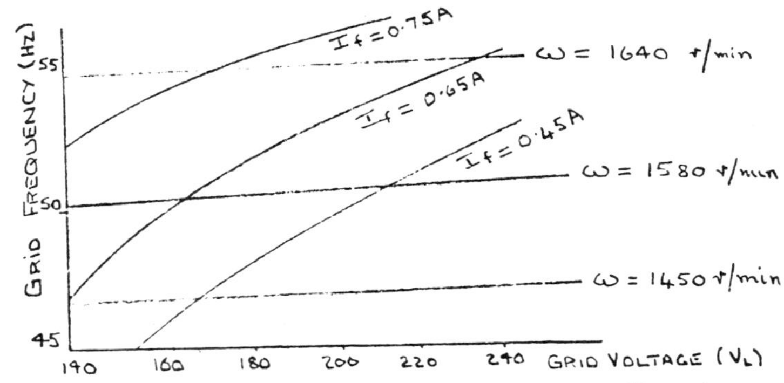

FIG. 10 DETERMINATION OF OPERATING POINT FOR A PARTICULAR LOAD DEMAND.

SOME PRACTICAL ASPECTS OF
SMALL WIND ENERGY CONVERSION SYSTEMS

G R Watson

Northumbrian Energy Workshop Ltd
Tanners Yard
Gilesgate
Hexham
Northumberland

Energy Centre Ltd
Pennine House
Osborne Terrace
Newcastle upon Tyne
NE2 1NE

Abstract

The quest for efficiency in small wind energy conversion systems often takes very different directions from those considered important by multi-megawatt designers. The system must be considered as a whole and the importance of rotor performance pales beside the other factors that will influence the actual usable energy output. The paper considers some of these factors, not all of which have technological solutions.

Resource Assessment

The potential SWECS owner has only considered the possibility because he believes the resource is available. The statement "there's always a good blow" is extremely difficult to quantify and individual perceptions of wind speed vary wildly. Users of wind energy up in Fair Isle (Force 8 exceeded 119 days in 1979) will rate "a good blow" at a somewhat higher wind velocity than a user, say, in Reading.

Monitoring must thus play an important role in whether a SWECS will be viable or not. This is why Northumbrian Energy Workshop have strained their limited resources to produce a special logger for the purpose. But the problem immediately arises with the SWECS "consumer" of immediacy. Western consumer society assumes instant availability and any deferrment means a loss of interest. The SWECS supplier then has the choice of losing the sale or installing in the hope that the site is as good as is claimed.

The compromise solution is short-term testing, say three months, and then correlating with Met Office. The ability of frequency spectrum analysis on the new logger makes this more of a possibility, but the main problem is the Met Office correlation, which, through published channels, could take another six months. It would seeem to be an early essential of any overall UK wind energy programme to increase the number of official stations, the gaps are enormous and the locations of dubious validity.

Even where direct comparisons are attempted, as in Figure 2, the whole validity of comparing local data with remote Met Office sites is called into question. The solid line is the weekly average wind speed (Casella wind-run) on Deadwater Fell, Northumberland during 1979 (when it could be read). A second wind-run unit was installed 40 miles away on a similar site at Moor House, Cumbria and used to adjust the Moor House (Met Office) 10 year mean. In this particular case, the information only vindicated a situation that was pretty well proven already (as can be seen from the 34 mph weekly average in September) but, in different circumstances, it may not always be possible to mount the second anemometer at the local Met Office site (and have it read weekly) - or, as is more probable, the nearest site is on the end of a pier and the SWECS user lives on the top of hill 100 miles away.

Demand Assessment

The only units supplied "off-the-shelf" are the little 200 Watt systems (system 1 of Figure 1) and the users of these are generally the most energy conscious! Defining Small Wind Energy Conversion Systems as units less than 100 kW rated covers a vast range of system variations and down at the 0.2 kW end is the mysterious world of low voltages and high currents. When cables are costing over £3/metre to reduce I^2R losses the turbine position is inevitably a balance between ideal wind and minimum cable length.

As output is fairly nominal then demand has to match and the client is always first asked to divide his requirements into "convenience" and "critical". Even with larger size systems, it is surprising how often an electric kettle or an automatic washing machine leaps from critical to convenience when the inverter cost is discussed.

It is not just a simple matter of specifying minimum energy equipment - fluorescent lights are super-efficient, but who can live with them? It is in this area of stand-alone units that the value of low wind performance is greatest. To a user dependent on wind energy as a sole source of electricity, then the few Watts generated in gentle breezes during the Summer is infinitely more valuable than the extra hundreds of Watts generated in Winter gales. This is where those machines with rotors that turn out of the wind can score, improving low wind performance by not having the rotor loading problems of high wind velocities.

Larger machines will rarely have this feature, but these will almost always be used forheating purposes (or be grid connected) and for these applications, the extra output in Winter winds is just what is required to match peak heating demands, whilst Summer performance is of marginal interest.

System Practicalities

The most telling formula in wind energy is that which multiplies all the individual component efficiencies to arrive at an overall wind energy in/user energy out efficiency. Slowly, the equipment is being developed to improve SWECS performance, inverters used to be a real problem area as a 5 kW inverter may be 85% efficient at full load but require 0.25 kW to do anything. The problem was exacerbated by the fact that a 1 HP induction motor may require 1100 Watts when running but will require 2500 Watts to start. The problem was solved by additional electronics, and the overall system efficiency (and battery storage requirement) improved immediately.

Like most areas of technology, wind energy is only just beginning to feel the potential of the mighty micro and it is certainly in the area of control that the biggest improvements in efficiency are likely to come about. Centralised control units for a wind/diesel system (system 2, Figure 1) already allow for diesel running periods to prevent injector problems, load sharing between priority loads and a fairly sophisticated battery charging procedure with heating ballast balance. If further developments can improve the power matching problem then we will see a far greater increase in overall system efficiency than can ever be achieved by rotor aerodynamics.

Where the more sophisticated rotor analyses will prove useful is in improving system reliability. Available systems to date have not convinced me of a thorough understanding of the loads involved by those responsible for detail design, involving Northumbrian Energy Workshop in considerable additional work to overcome detail faults. When a well-known manufacturer of many years standing has to withdraw a new unit because of hub failures then one has to wonder whether the split between university research and industry is becoming as wide in this industry as in any other. For such a new industry this would be a great pity.

SWECS operators can be split into two types, those purely interested in the end product (usually 240v 50 Hz) and those involved with the whole wonder of wind energy. Reliability problems with the latter users is not really a problem, they are generally prepared to correct the minor problems usually caused by water ingress. Those, often institutional, users interested in a power supply have to follow a maintenance schedule for in all cases, it has proven essential for aerogenerators to have regular inspections. A bearing which has the design life of the car it is fitted in could be worn out in a wind turbine within three months. Wind turbine manufacturers have, however, a lot to learn about ease of part replacement and minimising labour time.

The great unknown in SWECS operation at present is the effect of lightning. All the care and attention in the world will not prevent the damage that can (sometimes) by caused by a lightning strike. Presumably someone, somewhere, is doing research into this?

Economics

Consideration of the system costs in Figure 1 leads to the conclusion that the buyer will long since have deceased when the system breaks even. The fact, however, that a few struggling concerns exist manufacturing and installing these units show that it is not quite that simple. Systems 1 and 2 often require no economic justification, grid connection would be prohibitively expensive and delivered diesel cost plus the usual diesel problems make wind the most suitable energy source.

Other systems require a certain level of commitment by the purchaser, particularly as they are generally required to carry VAT, although stand-alone units do not. The site having the wind regime of Figure 3 would have a payback of 6 - 9 years depending on the degree of useful energy utilisation. This latter point is important as not all the projected energy output could be utilised and thus should not really be costed. The whole costing exercise is, in any case, overshadowed by the projected future inflation rate of energy.

The big advantage a SWECS will have is its cost/unit comparison with electricity at consumer purchase price, not CEGB producer cost. SWECS costs are also dropping, and will continue to do so as production batch sizes increase. The 200 Watt unit will be dropping by 20% this year (dependent on the £) and inverter prices have already fallen by 40%.

The question of capital cost is, however, paramount. Northumbrian Energy Workshop concentrates on refurbishing multi-blade water pumping units as a new unit with all its material and labour content will cost perhaps 20 times the price of a modern electric submersible pump. Even though the payback may appear feasible, the initial capital cost is daunting even to a farmer used to the price of combine harvesters. If the unit is to produce electricity, however, then an unreliable mains supply could tip the balance sufficiently to justify a SWECS.

For those with a mains supply (or prepared to live with the frustration of no wind power when the grid fails) then System 4 indicates the substantial reduction in capital cost if the SWECS is connected to the local board network. The question of SWECS grid connection not only deserves clarifying but should surely deserve support from the Electricity Authorities. The technical doubts revolve around a series of consumer tests on acceptable levels of light flicker that bear little relevance when the only person affected is benefitting from the cause of these minimal voltage changes. The approach undertaken in Denmark seems sensible and worthy of close study (Ref 1).

Conclusions

Every Small Wind Energy Conversion System is ideally different as the energy source changes from site to site. There will never be an ideal wind turbine for every site and each system becomes a balance between various factors.

One area that urgently needs attention by the UK Wind Community is that of agreed standards. At present, there is considerable dispute over even such basic decisions as a standard rated wind speed. Any more complex information is either not available or unquantified. All the machines installed to date by Northumbrian Energy Workshop have been well proven (ie, the faults are known and can be rectified) but none are manufactured in the UK. The latest Northumbrian Energy Workshop unit is American and its reputation is based on testing work carried out at the

US test station. A Danish machine will shortly be ordered, again because of adequate test data from the Danish test station. There is no way that a UK wind manufacturing base will ever be established without a similar facility being created. Not only could they not export, it is doubtful if they could even develop a sufficiently reliable machine for the UK market. A potential market would seem to exist in the UK for some 20,000 units, perhaps worth £150M. For such a market, it must be worthwhile enabling a UK manufacturing base. Export potential would of course be legion. There are 53,000 un-electrified villages in Mexico alone, and Mexico has petro-dollars.

With or without UK turbines, the most rapid change in the adoption of wind energy is likely to come about should the availability of grid connection be endorsed. 20,000 units could mean 200 MW of connected capacity at no cost for plant or maintenance to the grid.

Finally, we have still to gather support for arguments over planning restrictions. Northumbrian Energy Workshop have installed machines in two National Parks with no problems and have yet to experience real problems. But they will come, particularly as towers get taller. Ignoring the conservationist lobby means a perilous future for the widespread adoption of wind energy.

REFERENCES

(1) First Workshop on Methods and Standards for Wind Power Performance Testing, Copenhagen March 1980

– FIG 1 –

System 1

200 Watt remote system

	£	%
Wincharger 1222H 0.2kW wind turbine (super h.d. version)	460	30
10m hinged tower	465	30
1.5kWh battery store (12v 128Ah)	180	12
Electronic controller	75	5
Cabling (typ)	140	10
Installation (typ)	200	13

TOTAL COST £1520

System 2

3 kW high reliability tele-communications system

	£	%
North Wind HR2 3 kW wind turbine	4300	28
40 ft tower	2400	15
1.75 kVA auto-start diesel set	1600	10
24 kWh battery store (tub.plate) (24v 1000 Ah)	2500	17
(plante	5200)	
(nickel cad	6000)	
Electronic controllers and ballast loads	2500	17
Remote hill-top installation (typ)	2000	13

TOTAL COST £15300 - £18800

System 4

Comparison 4 kW stand-alone vs inverter grid connection

	£
Battery storage (36kWh)	2950
5 kW inverter	1650
3.75kW diesel set	820
Associated controls	350
	£5770
4 kW synchronous inverter	510
Associated controls	220
	730

System 3

10 kW domestic water heating system

	£	%
Elektro WVG 120D 10 kW turbine	4700	57
Controllers and sensors	1100	13
14.5m tower	1050	12
Misc. (lightning protn., heating elements)	610	8
Installation (typ)	800	10

TOTAL COST £8260

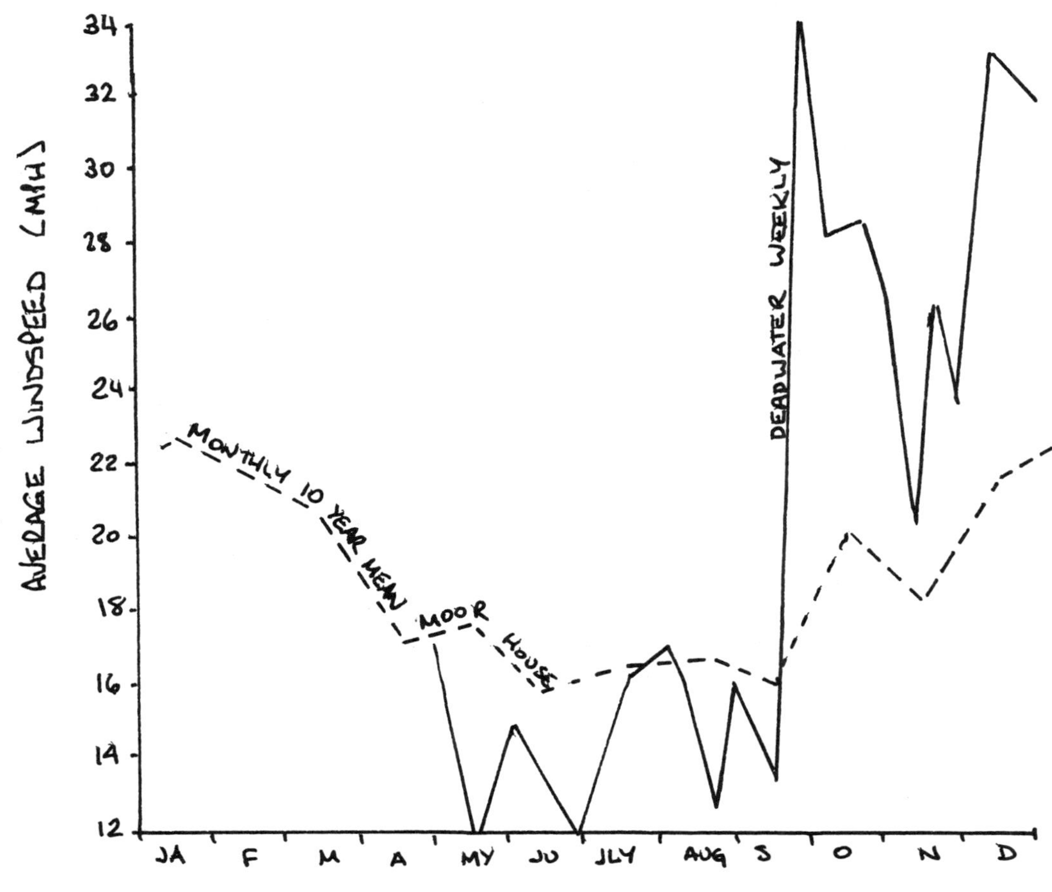

FIG 2 - DEADWATER & MOOR HOUSE WIND DATA 1979

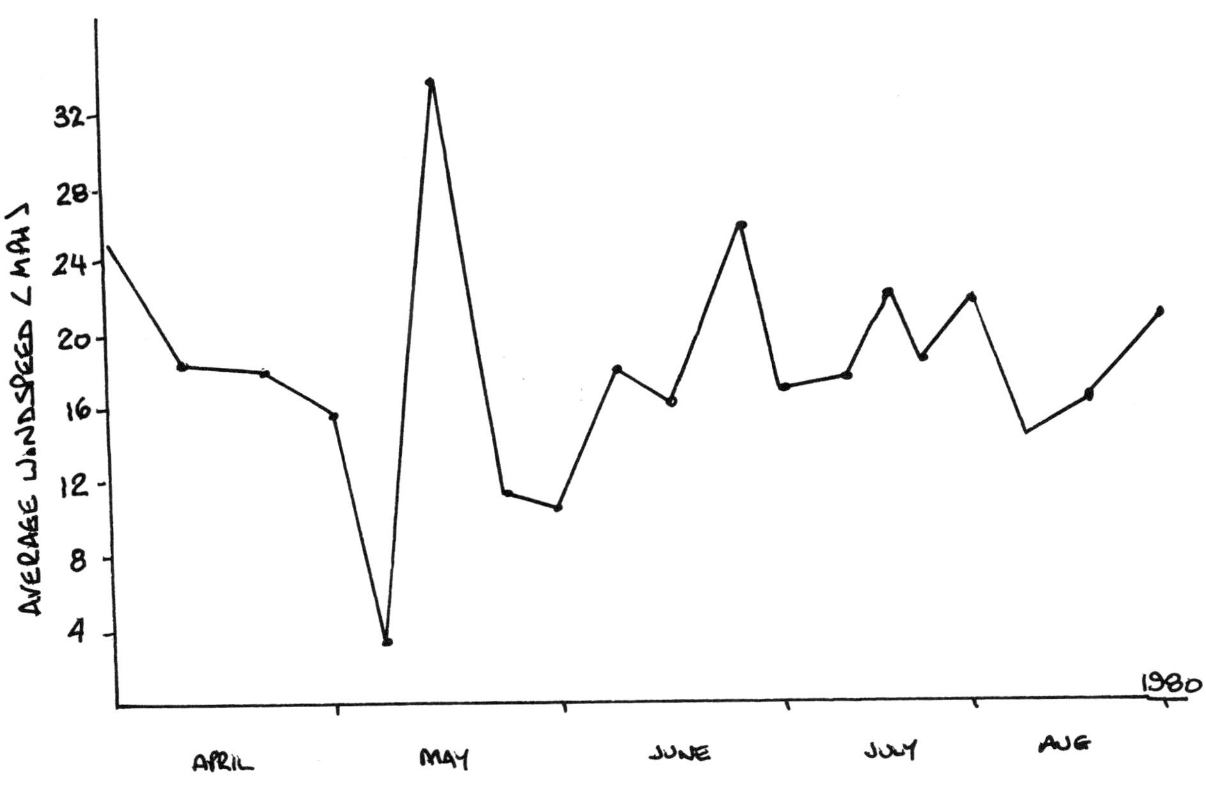

FIG 3 - POTENTIAL 55KW SWECS SITE WIND DATA

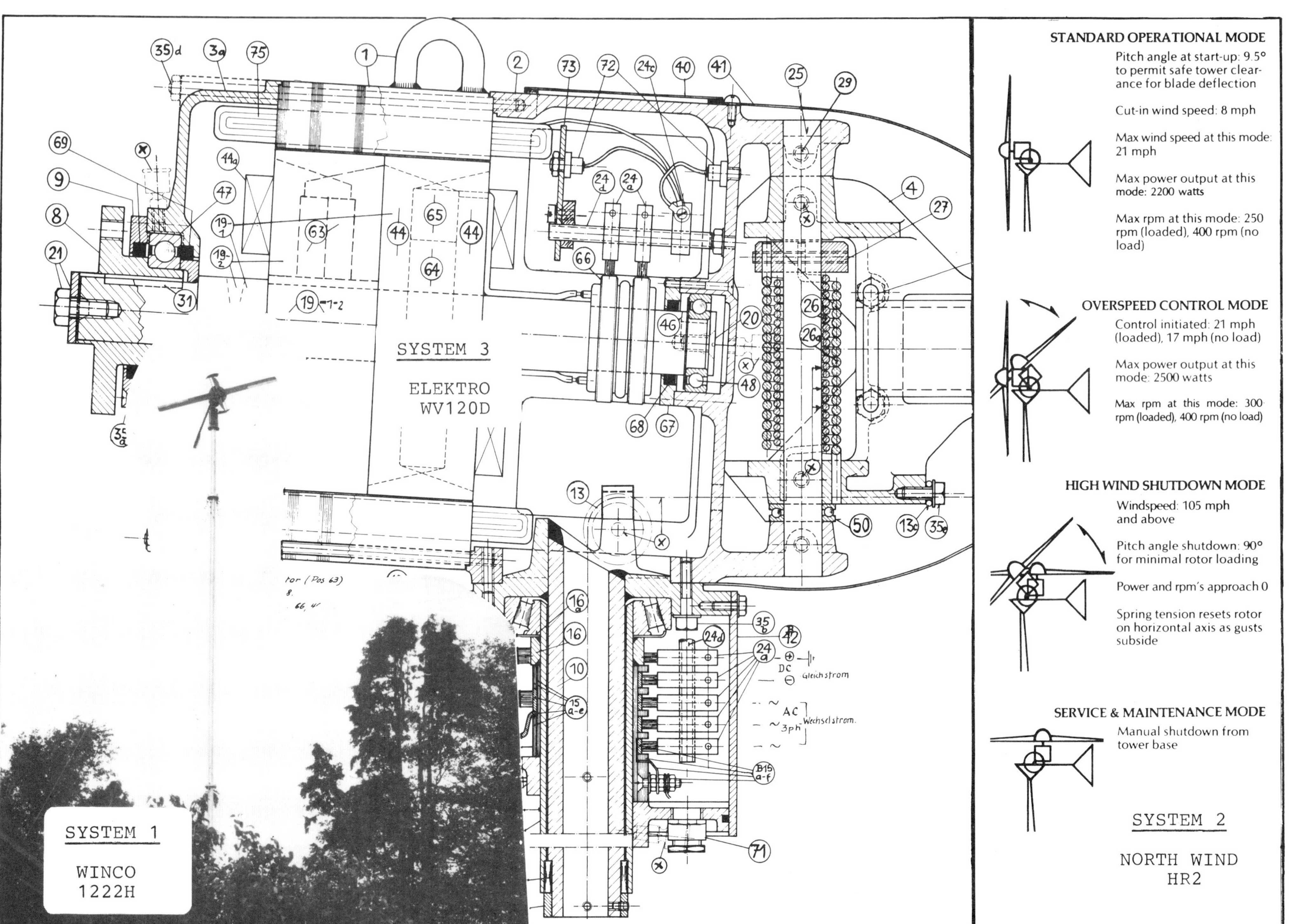
STANDARD OPERATIONAL MODE
Pitch angle at start-up: 9.5° to permit safe tower clearance for blade deflection
Cut-in wind speed: 8 mph
Max wind speed at this mode: 21 mph
Max power output at this mode: 2200 watts
Max rpm at this mode: 250 rpm (loaded), 400 rpm (no load)
OVERSPEED CONTROL MODE
Control initiated: 21 mph (loaded), 17 mph (no load)
Max power output at this mode: 2500 watts
Max rpm at this mode: 300 rpm (loaded), 400 rpm (no load)
HIGH WIND SHUTDOWN MODE
Windspeed: 105 mph and above
Pitch angle shutdown: 90° for minimal rotor loading
Power and rpm's approach 0
Spring tension resets rotor on horizontal axis as gusts subside
SERVICE & MAINTENANCE MODE
Manual shutdown from tower base
SYSTEM 2
NORTH WIND HR2
SYSTEM 3
ELEKTRO WV120D
SYSTEM 1
WINCO 1222H
DC
Gleichstrom
AC
3ph
Wechselstrom.

MEASUREMENT & INTERPRETATION OF WIND TURBINE WAKE DATA

D.J. Milborrow
Central Electricity Research Laboratories
Kelvin Avenue, Leatherhead, Surrey

ABSTRACT

A review of wind turbine wake data has shown that there is a need to define parameters to characterise wakes and their decay rates; this will enable comparisons to be made between the performance of rotors and simulators and also enable mathematical modelling techniques to be improved. It is suggested that thrust and power extraction coefficients based on the mean velocities in the wake may serve this purpose and the relationship between these and other performance parameters are examined. It is also shown that the apparent variation of the power extraction coefficient in a wake enables the turbulence variations to be inferred, without the need for sophisticated measurements. A consideration of the fundamental physics of wake generation leads to the conclusion that the use of simulators to reproduce wind turbine rotor wake characteristics is soundly based but may need to be re-examined quantitatively and this conclusion is reinforced by an analysis of available data.

NOMENCLATURE

A	Wind turbine or simulator disc area	m^2
Cp	Power coefficient, $\dfrac{\text{Power generated}}{1/2\ \rho A u^3}$	
Cp	Power extraction coefficient $\dfrac{\text{Power extracted, W}}{1/2\ \rho A u^3}$	
C_T	Thrust coefficient $\dfrac{T}{1/2\ \rho A u^2}$	
F	Mean axial velocity at disc plane	m/s
K	Drag coefficient based on local velocity $\dfrac{T}{1/2\ \rho A F^2}$	
p	Static pressures, relative to free-stream static pressure	N/m^2
p_1	Static pressure - upstream face of disc	N/m^2
p_1'	Static pressure - downstream face of disc	N/m^2
p_0	Total pressure, undisturbed stream $(= \frac{1}{2}\ \rho u^2)$	N/m^2
p_2	Total pressure in wake	N/m^2
Q	Volumetric flowrate	m^3/s
R	Radius of disc	m
T	Thrust	N
u	Free stream velocity	m/s
v	Velocity (axial) at any point	m/s
W	Air power extracted by wind turbine	W
ρ	Air density	kg/m^3

Refer also to Figure 2

INTRODUCTION

The interpretation of measurements made in wind turbine wakes is not a straightforward task. Such measurements are needed for studies of wind turbine cluster performance but can rarely be derived from full-size power generating machines. The cluster studies have therefore made use of models - experimental and mathematical - with wake characteristics derived from simulators and empirical wake decay data (Builtjes 1978). These model characteristics have been corroborated with data derived from wind tunnel tests of small rotors and with limited data from field tests of operational machines. However, allowances need to be made for the effects of scale, of ambient turbulence level and of the performance characteristics of the machine.

With the prospect of wake data from large power generating machines becoming available in the near future, methods of interpretation need to be specified to ensure that they can be correlated with existing data from smaller machines, simulators and mathematical models. Comparisons between datasets have generally been made by examining the wake centreline velocity deficits, wake velocity profiles and the decay rates of the former. Single parameters, which define the characteristics of a complete wake at any section, would clearly aid comparisons.

Studies of wake characteristics cannot be divorced from studies of aerodynamic performance. Two studies have recently examined the relationships between characteristics and performance parameters (Milborrow and Ainslie, 1980 and Sforza, 1980). Both these concluded that further basic studies of the fluid mechanics of the flow through porous obstacles - such as wind turbine rotors - and of wake characteristics were needed.

This study therefore examines the basic physics of flow through a wind turbine and continues with a review of the basic concepts associated with aerodynamic performance. Two parameters are identified which can be used to characterise both performance and wake properties - the thrust and power extraction coefficients - and the derivation of these from wake data is examined. The effects of parameters known to influence wake characteristics are also examined in terms of their effects on these coefficients.

OBJECTIVES

The objectives of this study are as follows:-

i) To review the basic concepts of aerodynamic performance theory for wind turbines and the fundamental physics of wake generation by porous obstacles.

ii) To specify suitable techniques for analysis of wake characteristics so as to facilitate comparisons between laboratory and field test data.

iii) To examine the relationships between wake characteristics and the aerodynamic performance of wind turbine rotors.

iv) To provide a basis for a review of the assumptions made in mathematical modelling techniques.

v) To re-examine the validity of using gauze simulators for studies of the performance of arrays of wind turbines.

PHYSICS OF WIND TURBINE WAKES

Although there have been a number of wake studies which have examined the flow characteristics produced by wind turbine rotors and simulators, usually formed of wire gauze, the mechanisms producing the flow patterns have received less attention.

The concept of a wind turbine rotor being equivalent to a porous obstacle in an airflow can be justified by the laws of thermodynamics (Milborrow, 1980) and a more detailed consideration of the flow confirms this view. Porosity is to some extent a misleading concept since both wind turbine rotors and gauzes comprise assemblies of bluff objects – rotating blades and small wires respectively – whose influence on the local flow may be expected to be quantitatively similar. Both blades and wires accelerate the flow locally, as it passes round the obstacle, and introduce aerodynamic drag which is partly associated with surface friction, partly with pressure differences and partly with turbulence generated in regions of high shear and/or separation. The surface friction is generally small and results in the conversion of kinetic energy into heat. Since the local accelerations in the flow around the blades or wires generate velocity and pressure gradients in the plane of the rotor, further turbulence is generated in the mixing process as the flow moves away from the rotor.

In making comparisons between rotor wakes and those of static objects certain difficulties must be noted. The rotation of a blade will clearly introduce periodicity into a rotor wake although there is evidence that this disappears after about five diameters (Vermeulen, 1978). The associated swirl is unlikely to be dissipated but is known to be small. Moreover, the local flow accelerations produced by the aerofoil section of a rotor blade are likely to be much higher than those produced by simulator wires. The larger velocity gradients in a rotor wake are likely to be dissipated more rapidly, however, since the relative velocities with respect to moving rotor blades are higher than those around simulator wires; the Reynolds Number, based on any characteristic length, will therefore be higher.

The characteristics of wind turbine rotor wakes, and their subsequent decay, have been the subject of a number of studies (e.g. Vermeulen, 1980). Other studies have confirmed that rotor characteristics are similar to those of simulators (e.g. Builtjes, 1978). Little attention has been paid, however, to the energy exchange processes in the flow - from the mean flow to turbulence and heat - and it is therefore proposed to consider these in more detail making use of standard parameters used to characterise wind turbine performance.

AERODYNAMIC PERFORMANCE - BASIC CONCEPTS

Three parameters are central to an understanding of the aerodynamic performance of wind turbine rotors - the thrust coefficient, C_T, the ratio between the mean disc velocity and the freestream velocity, $F/\bar{u}$, and the power coefficient, Cp. As the latter is a function of the geometry and size of specific rotors, a more fundamental parameter, the 'power extraction coefficient', C_P, will be used for this study. Algorithms are included in the Appendix for the calculation of thrust and power extraction coefficients from wake traverse data. These omit the contributions due to the turbulent momentum flux, since the information needed to integrate the turbulence terms, i.e. the radial distribution of intensity of each component, is rarely available. Although it is important to establish the levels of turbulence intensity in wind turbine wakes, variations in the total turbulent momentum and energy fluxes can be derived from an examination of the changes in fluxes calculated using the mean velocities and pressures. Any apparent reduction of energy based on mean values must result in an increase of turbulent energy. Detailed theoretical and experimental studies of the energy exchange processes in wakes of solid objects have been made by Carmody (1964) and Chevray (1968), to which reference may be made for formulation of the additional terms in the thrust and energy equations.

Thrust Coefficient

The standard method of calculating the thrust coefficient from wake traverse data is included in the Appendix. However, this method is difficult to apply to field test data, as measurements are required at radii greater than that of the rotor disc. Since accuracy and economy are likely to be served by deploying instruments over as small an area as possible, an alternative method of calculation, based on the flow passing through the machine, is derived in the Appendix. This approach requires an estimate to be made of the velocity at the plane of the rotor disc and should therefore be used in conjunction with measurements made close to the rotor.

Power Extraction Coefficient

Comparatively little attention has been paid in wind turbine aerodynamic studies to the power extraction coefficient and frequently no distinction is drawn between power generated and power extracted. Insufficient emphasis, for example, is placed on the fact that the classical 'Betz limit' is a power extraction limit. The definition of power extraction coefficient is therefore very similar to that of a power coefficient and the ratio of the two coefficients C_p/C_P denotes the true aerodynamic efficiency of a wind turbine rotor. (The power coefficient is not a true measure of efficiency, nor is the ratio of the power coefficient to the Betz limit although both definitions are loosely and inaccurately used). It may be anticipated that the power extraction coefficient is of more significance in examining wake characteristics than the power (generation) coefficient and methods of calculation are therefore included in the Appendix.

REVIEW OF AVAILABLE DATA

Thrust Coefficients

Derivation of rotor thrust coefficients enables the total loadings on the rotor shaft and turbine support tower to be determined and estimates of blade loadings to be made. Although these can be derived from performance prediction techniques, confirmatory data are required. The behaviour of the thrust coefficient at high rotor loadings is also important, but a matter of dispute, as discussed by Milborrow and Ainslie (1980), who advanced the hypothesis that it increases continuously with rotor loading, in contradiction to the simple theory. A study of the relationship between the disc velocity ratio, F/u, and the thrust coefficient confirms the conclusions drawn from this study. Figure 1 shows the relationship derived from simple momentum theory, together with experimental data from a number of tests of porous discus. It is clear that the simple theory does not predict the velocity ratio correctly at high loadings and that thrust coefficients greater than unity can be achieved; no data from rotors are plotted but Dugundji et al (1978) have reported rotor thrust coefficients exceeding unity in laboratory tests.

A number of authors have reported satisfactory agreement between thrust measurements and estimates derived from integration of the velocity and pressure profiles in the wake of an obstacle. These include Carmody (1964), for a disc, Chevray (1968), for a streamlined body, and Vermeulen (1978), for a wind turbine rotor and a simulator. The latter showed that omission of the turbulent momentum flux terms did not lead to large errors. The integration of the wake parameters may therefore be carried out at any convenient plane, provided that the pressure term is not neglected close to the rotor (i.e. at distances less than 3-4 diameters.

Data suitable for testing the accuracy of the alternative (streamline) method of calculation are scarce but some measurements made on wire gauze screens at the University College of Wales, Swansea, kindly supplied by Mr. R.T. Griffiths (private communication) have been used to confirm the validity of the method. The gauze screen had a drag coefficient of about 2.2; measured velocity and pressure profiles are shown in Figure 2. The thrust coefficient, derived by the standard method, at the station 1.85 diameters downstream of the rotor was 1.01, while the streamline method at the station 0.28 diameters downstream of the rotor yielded a value of 1.03.

The chief difficulty, for the present, with the streamline method is determining the appropriate velocity at the rotor plane. Although the streamline curvature program enables estimates to be made, these appear, as shown in Figure 1, to be lower than those derived from (somewhat sparse) experimental data; however, the latter do appear to show a consistent trend. As the centre line velocity deficit of a wake tends to be used to characterise its properties, the minimum values of the centre-line velocity ratio (1 - deficit) are also plotted in Figure 1. Again, there is a weak correlation evident, suggesting lower values than those predicted by the streamline curvature studies. These minimum values of centre line velocity tend to be reached about 1 diameter downstream of the rotor and to remain constant for 2-4 diameters.

Power Extraction Coefficients

There is a lack of data on power extraction coefficients but calculations, using techniques similar to those applied to the prediction of performance of axial flow fans (Milborrow, 1978), yield curves imilar to those presented for conventional power coefficients. A limiting value of 0.47 for the power extraction coefficient was derived from the streamline curvature analysis (Milborrow and Ainslie, 1980), but the present study, drawing on the experimental data of Figure 1, indicates that this may be pessimistic. It can easily be shown that the power extraction coefficient, $C_P = C_T.F/u$ and Figure 3 has been derived using this relationship and the experimental data of Figure 1.

Although the concepts of rotor power extraction coefficient and aerodynamic efficiency are useful when comparing the capabilities of rotors, exact values, including the determination of the limits, are likely to prove difficult to determine. As the flow moves past or through an obstruction, porous disc or rotor, there is a rapid conversion of energy from the mean flow to turbulence, as discussed in the Section on physics of wind turbine wakes. Calculations of 'apparent' power extraction coefficients, based on mean values of velocity and pressure, close to obstacles indicate that this conversion process can take place rapidly; for example, integration 0.18 diameters behind the gauze tested at the University College of Wales yielded an extraction coefficient of 0.73; at 1.85 diameters this had risen to 0.76. A study of available data from other sources, shown in Figure 4 confirms that 'apparent' power extraction coefficients based on mean values rise to high values which persist far downstream. These data emphasise the importance of turbulence generation in a wind turbine wake, since the power converted into turbulent energy can exceed that extracted by the machine. This observation may also explain the anomaly noted in the previous section where it was found that mean disc velocities appeared to be higher than predicted theoretically (Figure 1 refers). The discrepancy may be due to the rapid energy conversion from mean to turbulent flow, so that, at least for porous discs, power extraction appears as an increase of turbulence as well as a reduction of velocity.

All the curves shown in Figure 4 indicate that the 'apparent' energy extraction rises to a peak - between 1 and 6 diameters from the obstacle - and then falls. This implies that the turbulence intensity rises to a peak and then falls and is consistent with measurements made by Vermeulen (1978). Values of turbulence intensity, measured on the wake centre line behind a horizontal axis rotor have been included for comparison and show that the peak occurs at the same location, about 3½ diameters downstream of the rotor. The fall of turbulence intensity beyond this point is normally ascribed to dissipation into increased internal energy, i.e. heat, but the corresponding fall in the power extraction coefficients implies that some of the turbulent energy is recovered by the mean flow. This has been ascribed by Raj and Lakshminarayana (1976) to gradients in the turbulence quantities exceeding those in the mean flow.

Although the comparisons of apparent power extraction characteristics for rotors and simulators, shown in Figure 4, indicate satisfactory agreement, further data are needed on their turbulence charactersitics and on the effect of changes in the ambient turbulence level. The latter appears to have a marked effect on simulator wake behaviour as shown by the comparison included in Figure 4.

Effects of Scale

Although scale effects have a significant influence on the power generated by wind turbine rotors, their effects on power extraction are less marked. Estimates made using a momentum analysis, based on the standared techniques employed for axial flow fans (Milborrow 1978), indicate negligible changes of power extraction as the drag/lift ratio varies from 0 to 0.1. It was found that the power loss due to drag rose to compensate for the reduction of shaft power, as shown in Table 1 below. The rotor geometry used for the calculations was that of the Aldborough wind turbine (Nickols and Milborrow, 1981).

Table 1

Power Generation and Extraction of Aldborough Turbine

Tip speed/wind speed ratio 7.3

Drag/lift ratio, C_D/C_L	0.0	0.01	0.03	0.10
Power coefficient, Cp	0.54	0.51	0.44	0.19
Power extraction coefficient, C_P	0.55	0.55	0.56	0.56
Efficiency	0.984	0.918	0.789	0.341
- loss due to swirl	0.016	0.015	0.012	0.006
- loss due to drag	0.0	0.067	0.199	0.653

CONCLUSIONS

The principal conclusion derived from this study is that there is a need for further fundamental and systematic studies of aerodynamic performance and wake generation and decay for wind turbine rotors and simulators. The effects of rotor characteristics, ambient turbulence level and scale also need to be studied. Nevertheless, it is possible to use the limited data available to conclude that

i) The conventional formulae, derived from simple momentum theory, to define the relationships between rotor loading, flow retardation, power extraction coefficient and thrust coefficient are inadequate. The alternative correlations, based on data from porous obstacles which have been quoted in this study, need confirmation by rotor performance data.

ii) Thrust and power coefficients based on mean values measured in rotor wakes can be used to characterise wakes, since they are sufficiently sensitive to provide a sound basis for examining the influence of rotor type, rotor thrust and ambient turbulence level. Calculations based on field test data should be possible, if analyses based on the streamline locations are used.

iii) The generation of turbulence both at and downstream of a rotor or simulator strongly influences the values of performance parameters (particularly power extraction coefficients) derived using mean values measured in a wake. However, there is an energy exchange between the mean flow and the turbulence and, hence, turbulence generation and decay can be inferred from a study of the mean values of momentum and energy flux. Further studies of the relationship between the total apparent momentum and energy flux extracted, based on mean values, and the turbulence intensity may reduce the need for sophisticated fast-response instrumentation.

iv) The rates of energy exchange from mean flow to turbulence may differ between rotors and simulators of the resistive type used for wind tunnel tests. However, use of the latter is not invalidated but criteria other than equality of thrust coefficients used hitherto may need to be defined.

v) The effects of scale on the generation of rotor wakes are small. This conclusion follows from performance calculations and also from a consideration of the basic physics of the flow: the perturbations are similar whether produced primarily by lift or primarily by drag. It may therefore be inferred that variations in the drag/lift ratio will not affect the wake characteristics although further data are needed to support this conclusion, since wake decay rates may be affected by the differing turbulence characteristics.

REFERENCES

Builtjes, P.J.H., 1978, The interaction of windmill wakes, 2nd Intl. Symp. on Wind Energy Systs., Amsterdam, BHRA Cranfield

Carmody, T., 1964, Establishment of the wake behind a disk, ASME Trans., Jnl. Basic Engng. 87, 869-882.

Chevray, R., 1968, The turublent wake of a body of revolution, ASME Trans., Jnl. Basic Engng, 90, 275-284.

Dugundji, J., Larrabee, E.E. and Bauer, P.H., 1978, Experimental investigation of a horizontal axis wind turbine, U.S. Dept. Energy, Report COO-4131-T1.

Hoerner, S.F., 1958, Fluid Dynamic Drag, Author.

Milborrow, D.J., 1978, Performance prediction techniques for horizontal axis wind turbines, Wind Engng. 2, 3, 165-175.

Milborrow, D.J., 1980, The performance of arrays of wind turbines, Jnl. Indust. Aero. 5, 403-30.

Milborrow, D.J. and Ainslie, J.F., 1980, Calculation of the flow patterns and performance of wind turbines using streamline curvature methods, 2nd BWEA wind energy workshop, Cranfield. Multi-Science.

Nickols, W.R. and Milborrow, D.J., 1981, Development of the Aldborough wind turbine, Proc. IEE Future Energy Concepts Conf., London.

Raj, R. and Lakshrinarayana, B., 1976, Three dimensional characteristics of turbulent wakes behind rotors of axial flow turbomachinery, Trans. ASME (Jnl. Engng. for Power), 98, 218-226.

Sachs, P., 1978, Wind forces in engineering, Pergamon.

Sforza, P.M., Sherrin, P. and Smorto, M., 1980, Further studies on wind turbine generator wakes, AAIA/SER1 Wind Energy Conference, Boulder, Colorado.

Taylor, G.I., 1936, The aerodynamics of porous sheets, ARC R&M 2237.

Vermeulen, P., 1978, A wind tunnel study of the wake of a horizontal axis wind turbine, TNO (Netherlands) Report 78-09674.

Vermeulen, P., 1980, An experimental analysis of wind turbine wakes, 3rd Intl. Symp. on wind energy systs., Copenhagen, BHRA Cranfield.

ACKNOWLEDGMENT

This work was carried out at the Central Electricity Research Laboratories and is published by permission of the Central Electricity Generating Board.

APPENDIX

CALCULATIONS OF THRUST AND POWER EXTRACTION COEFFICIENTS

THRUST COEFFICIENT

Standard Method

A control volume is specified, of radius S (S>R). The analysis proceeds on the basis that:-

Rotor thrust = Rate of change of momentum
= Momentum flux entering control volume
- momentum flux leaving control volume axially
- momentum flux lost to control volume radially

whence

$$\frac{T}{1/2\rho\pi R^2u^2} = C_T = 4\int_0^{S/R}\left[\frac{v}{u}\left(1-\frac{v}{u}\right)-\frac{1}{2}\frac{p}{\frac{1}{2}\rho}u^2\right]\frac{r}{R}\,d\frac{r}{R} \qquad \ldots \text{(A.1)}$$

In practice mixing of the airstream which has passed through the rotor with the adjacent streamlines necessitates the use of a control volume with as large a radius as possible. Otherwise the assumption that the airflow leaving the control volume radially retains the momentum flux of the undisturbed stream becomes invalid.

Streamline Method

The thrust of the rotor may be derived from an integration of the static pressure differences across the disc.

Hence the thrust of the rotor is given by:-

$$T = 2\pi\int_0^R (p_1 - p_1')r\,dr \qquad \ldots \text{(A.2)}$$

Provided there is no mixing and no secondary losses downstream of the rotor and writing Bernoulli's equation for streamlines

$$C_T = 2\int_0^R\left(\frac{P_0 - P_2}{1/2\rho u^2}\right)\frac{r}{R}\,d\frac{r}{R} \qquad \ldots \text{(A.3)}$$

It is important to note that the radii used in this equation apply at the rotor disc and are associated with pressures and velocities on the corresponding streamlines in the wake.

To counteract errors due to mixing at the edges of the rotor wake, the limits of the integral may be extended. Equation (A.2) remains valid since any terms at r>R would equal zero at the plane of the rotor.

POWER EXTRACTION COEFFICIENT

"Control Box" method

By specifying a notional cylindrical control volume the power extraction coefficient may be determined using a procedure similar to that employed for thrust coefficient in the Section on Standard Method. This gives

$$C_P = 2\int_0^{S/R} \frac{v}{u}\left(1 - \frac{p}{1/2\rho u^2} - \frac{v^2}{u^2}\right)\frac{r}{R}\, d\frac{r}{R}$$

The weaknesses of this method are similar to those discussed in the Section on Standard Method, i.e. the control volume must be large to ensure accuracy on account of mixing.

Streamline Method

There are a number of ways of formulating algorithms for calculations of power extraction on a streamline basis. The basic equation is

Power extraction = Upstream power flux - downstream power flux

hence

$$C_P = \frac{dQ}{\pi R^2 u}\sum\left(1 - \frac{p}{1/2\rho v^2} - \frac{v^2}{u^2}\right) \qquad \begin{matrix} r/R = 1 \\ \\ r/R = 0 \end{matrix}$$

In practice, it is advisable to extend the limits of integration beyond the rotor disc plane to account for mixing. Provided that the changes in streamline radii are calculated between upstream and downstream measuring stations, this procedure is quite valid as power is a scalar quantity and it is a difference which is being derived.

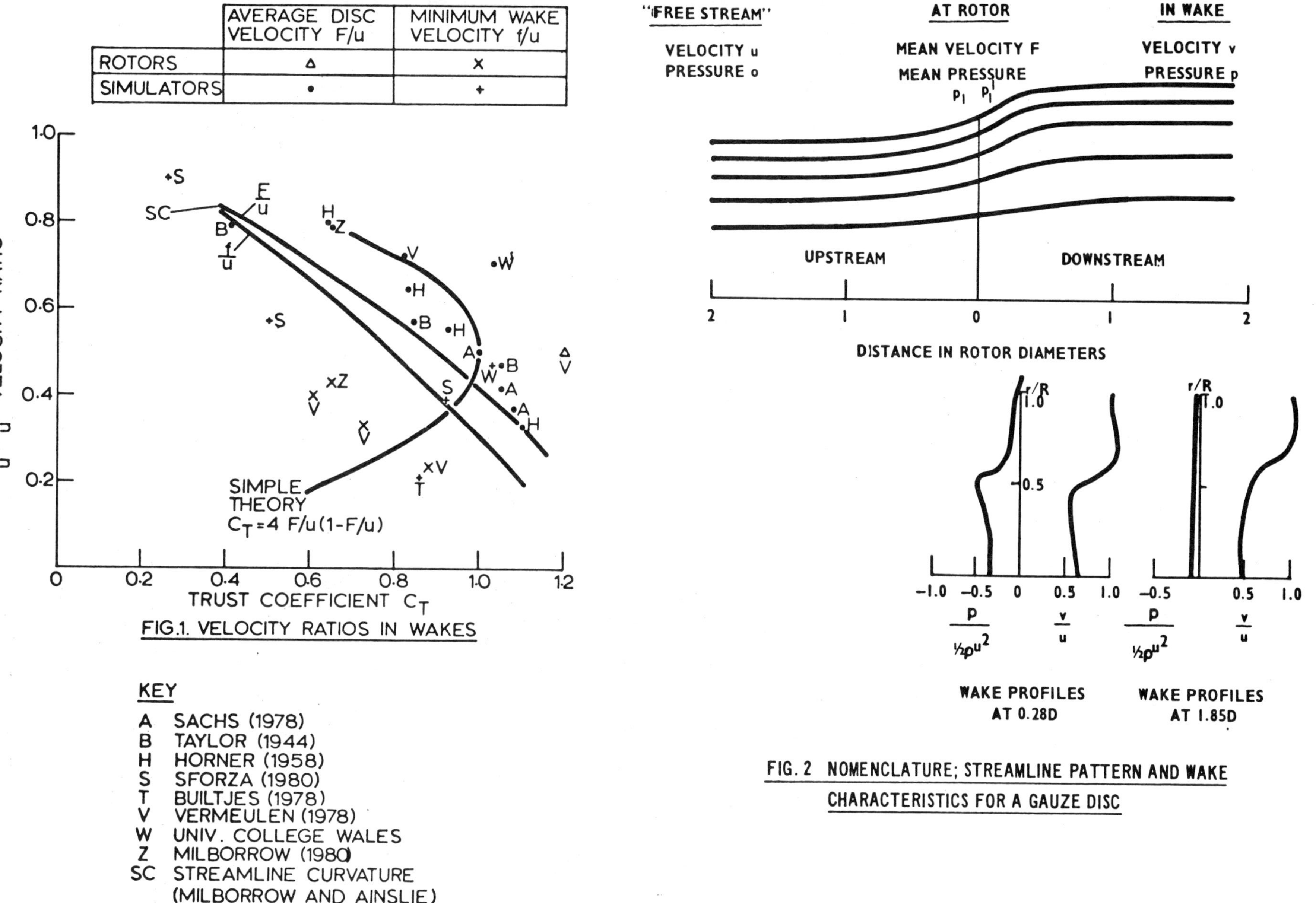

FIG.1. VELOCITY RATIOS IN WAKES

DJM/CPR(20.3.81)RL 3.3.5509

FIG. 2 NOMENCLATURE; STREAMLINE PATTERN AND WAKE CHARACTERISTICS FOR A GAUZE DISC

DJM/CPR(20.3.81)RL 3.3.5510

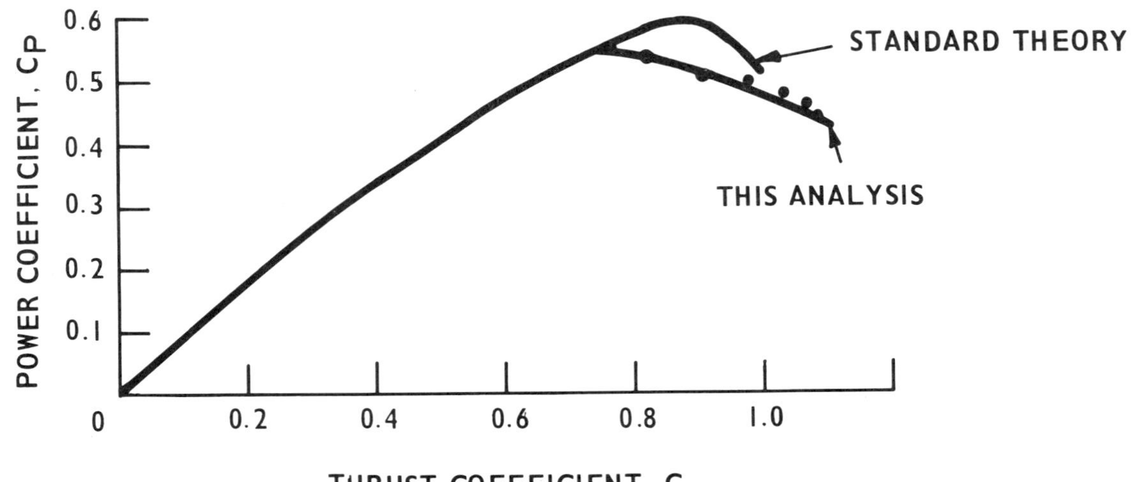

POWER COEFFICIENT, C_P
0.6
0.5
0.4
0.3
0.2
0.1
0
0.2
0.4
0.6
0.8
1.0
STANDARD THEORY
THIS ANALYSIS
THRUST COEFFICIENT, C_T

FIG. 3 RELATIONSHIP BETWEEN THRUST AND POWER

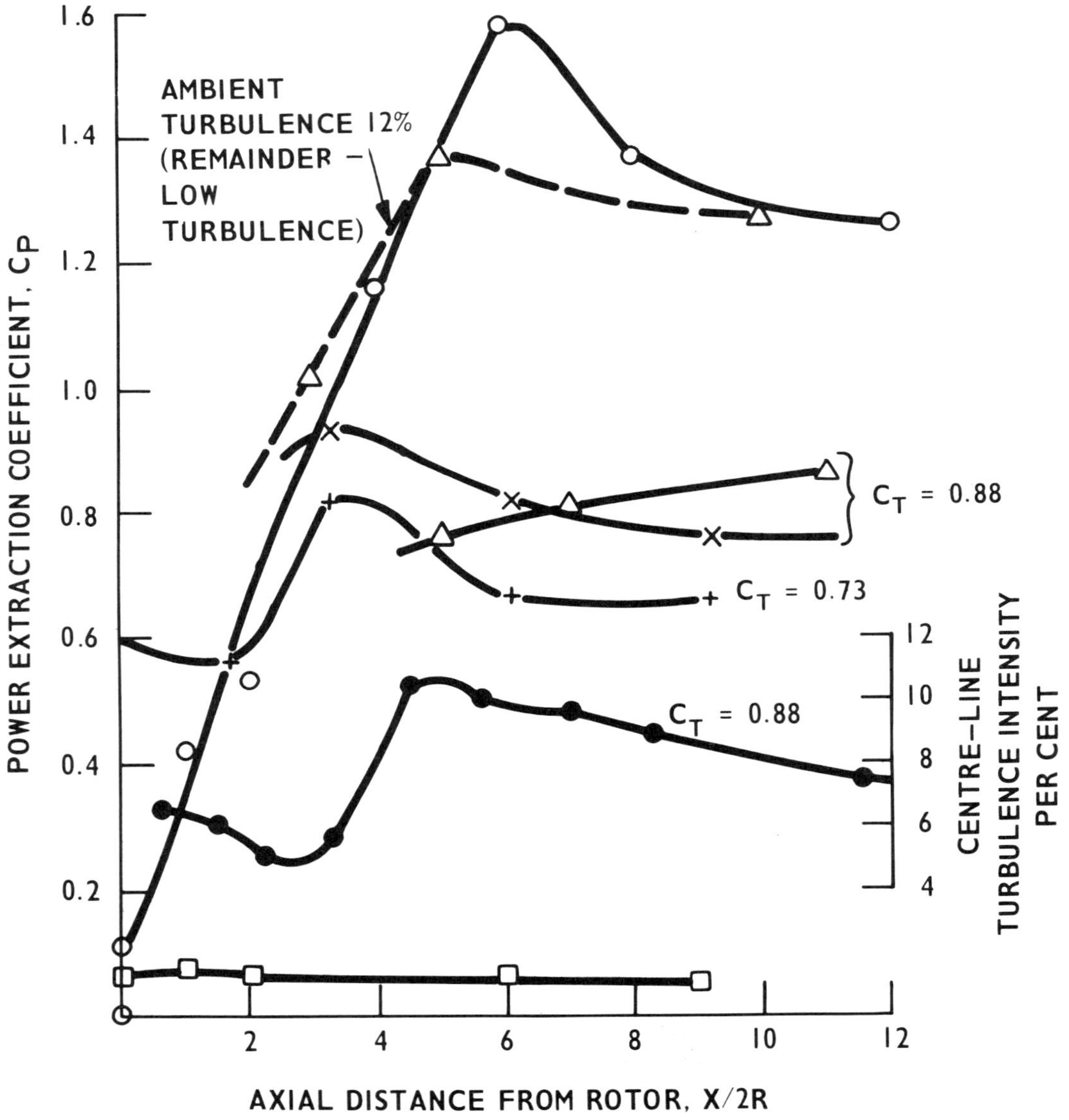

POWER EXTRACTION COEFFICIENT, C_P
1.6
1.4
1.2
1.0
0.8
0.6
0.4
0.2
AMBIENT TURBULENCE 12% (REMAINDER – LOW TURBULENCE)
C_T = 0.88
C_T = 0.73
C_T = 0.88
12
10
8
6
4
CENTRE–LINE TURBULENCE INTENSITY PER CENT
2
4
6
8
10
12
AXIAL DISTANCE FROM ROTOR, X/2R

FIG. 4 VARIATIONS OF POWER COEFFICIENT IN WAKES

- ○ CARMODY (1962) – DISC
- □ CHEVRAY (1962) – STREAMLINED BODY
- △ BUILTJES (1975) – SIMULATOR
- ×+ VERMEULEN (1978) – ROTOR
- ● VERMEULEN (1978) – TURBULENCE

THE EFFECT OF FIXED PITCH OFFSET ON A HIGH SOLIDITY VERTICAL AXIS WINDMILL

G. Stacey and P.J. Musgrove
Department of Engineering
Reading University.

Introduction

Increasingly interest in vertical axis wind turbines is directed towards large machines designed to produce electrical power either for grid connection or to service an isolated community. However development of small turbines for stand alone applications continues in many university groups and small companies.

Whereas the tendency in large turbines is towards low solidity, high tip speed ratio machines, the ability to self start is often of more importance in stand alone applications than the ultimate coefficient of performance.

This paper describes initial field tests carried out at Reading University on a high solidity turbine and the effect of pitch offset on the blades. Modifications to the multiple streamtube model are advanced which attempt to model the effects of flow curvature.

Wind Turbine Details

The turbine tested is a straight bladed vertical axis wind turbine with the following parameters.

Diameter	3 metres
No. of blades	4
Blade chord	0.4 metres
Blade span	1.25 metres
Blade profile	NACA 0015
Aspect ratio	3.125
Solidity	0.53

The blades are held vertically and are not allowed to incline.

Field Testing

The unpredictability and unsteadiness of the natural wind makes field testing an uncertain business. The performance results reported in this paper were obtained by the acceleration test method first described by Sharpe (Ref. 1) where the turbine is allowed to accelerate against its own inertia.

Measurements were taken of rotational speed, by means of a D.C. tachogenerator, and of the windspeed. Two anemometers were used placed approximately 5 metres on either side of the turbine; these were Gill anemometers fitted with Casella cups having a response length of 2.5 metres.

The three output signals were stored on an F.M. taperecorder for later analysis. During the tests the turbine was repeatedly braked to rest and then allowed to accelerate to its runaway speed. To date tests have been carried out with the blades mounted at right angles to the crossarm and with a pitch offset of $3\frac{1}{2}$ degrees (nose outwards).

Data Analysis

The data was first digitised by replaying the tape recorder through digitising equipment. Instantaneous values of the three channels were taken every second throughout the test period and transferred to the university's main-frame computer via punched tape. Data taken while the turbine was being braked was eliminated to leave approximately 1 hours worth of useable results.

The accelerator technique demands a notionally constant windspeed for a period during which the turbine is allowed to speed up. From this aspect the Reading Test Site is far from ideal with several close buildings leading to turbulence. Various constraints were tried in connection with the data to isolate periods of "steady" wind. The data was scanned to look for periods of 1 second where the windspeed on both anemometers at the beginning and end of the second were within a given tolerence. Obviously the tighter this constraint the more confidence can be placed on the result but the amount of resulting data diminishes. Eventually the criterion selected was that the two pairs of windspeeds should be within 10% of the mean of the four values. In addition periods where the windspeed dropped below 2 m/s were discarded due to the increased uncertainty in the turbines incident windspeed. This resulted in about 300 useable seconds of data for the blades square to the crossarm and 100 seconds when the blades were nosed out; i.e. 300 and 100 performance data points respectively.

The Tip Speed Ratio and the Power Coefficient were then calculated for these seconds of data and binned in tip speed ratio intervals of 0.25. Average values of Cp and Beta were calculated for each bin and the standard deviation in Cp calculated. These are shown on the performance curve where the error bar represents ± 1 standard deviation.

Note

$$\text{Power Coefficient } C_p = \frac{\text{POWER OUT}}{\text{POWER IN}} = \frac{I\,\omega\,\dot{\omega}}{\frac{1}{2}\rho A V^3}$$

$$\text{Tip Speed Ratio } \beta = \frac{\omega r}{V}$$

where I = Moment of Inertia of turbine
ω = Rotational Speed
A = Swept Area
ρ = density of air = 1.2
V = mean wind speed
r = Turbine radius

Results

The performance curves (Fig. 1) shown illustrate the large scatter obtained in results. This is primarily due to uncertainty remaining in the windspeed and a small ripple on the shaft tachogenerator which lead to significant uncertainty in calculating acceleration.

Optimum performance occurred in both cases at a tip speed ratio of about 1.85. With the blades mounted perpendicular to the crossarm a peak Cp of 0.09 was obtained. Pitching out the blades by 3½ degrees improved the performance to a peak Cp of 0.165.

In both cases the turbine self started and rapidly climbed into its design operating range.

Theoretical Aspects

The multiple streamtube approach conceived by Strickland (Ref. 2) for the Darrieus rotor and developed by several authors to apply to straight bladed vertical axis wind turbines can easily be amended to allow for pitch offset.

This model presumes inherent symmetry between the upwind and downwind power extraction "passes" of the blade so that a fixed pitch offset will increase the angle of attack at one pass and reduce it at the other. Consequently as one blade begins to stall the other has yet to reach optimum angle of attack and the overall peak performance diminishes. Coincidentally the model will predict identical performance curves for equal amounts of nose-in and nose-out. The effect of feeding various amounts of pitch offset into the model for this geometry is shown in Fig. 2.

Walters et al (Ref. 3) have discussed the effects of flow curvature leading to virtual camber and incidence on turbines with high chord to radius rotors. They have assumed an infinite tip speed ratio which allows calculation of a constant virtual camber and incidence around a revolution. They suggest that the lift and drag characteristics for this virtual aerofoil geometry should be used instead of that for a symmetrical aerofoil. However, investigation of the amount of virtual camber and incidence at the low tip speed ratios at which these high solidity turbines operate has shown that not only does the amount of camber and incidence change with TSR but also that it changes significantly within the revolution of the turbine. A programme was developed to carry out the conformal mapping needed to calculate these effective changes to the aerofoil shape. The programme results in a new camber line corresponding to the relative velocities along the aerofoil being brought parallel.

This is shown schematically in Fig. 3. Values of vertical camber and incidence at points around a revolution at appropriate TSR's are shown in Fig. 4.

Several methods were considered in order to see how to incorporate this effect into the authors current multiple streamtube model including aspect ratio effects (Ref. 4). The method suggested by Walters was only viable if a consistent set of data covering a range of cambered aerofoils could be obtained, but a literature survey failed to discover any such data sets.

Abbott and Von Doenhoff (Ref. 5) however had published a consistent set of data for a series of cambered aerofoils based on the NACA 0012 profile. They showed that in the unstalled region the effect of camber was broadly that of an increment in pitch of an uncambered aerofoil. Moreover for the range of aerofoils considered a 1% camber increment had much the same effect on lift and drag as a 1° increase in angle of attack.

It was therefore decided to include the virtual aerofoil effect into the multiple streamtube model by calculating the effective incidence and camber at each point around the revolution. Camber is converted to incidence on the 1% camber equals 1° incidence basis. These changes are then combined with the true angle of attack and the new incidence used to calculate lift and drag of the NACA 0015 aerofoils.

The authors acknowledge the speculative nature of this transformation but the results should offer considerable insight into the effects of the virtual camber and incidence. It has been shown that the effective camber and incidence both contribute to an effective nose-in at all points around the revolution.

Model Results

As expected with effective nose-in of several degrees the predicted performance of the turbine with no blade offset is poor. Various amounts of nose-out were included to counteract the virtual nose-in an optimum nose-out of 5° was predicted.

The results of the model are shown in Fig. 5 for a range of pitch offets including the $3\frac{1}{2}^{\circ}$ tried experimently.

Values of optimum Cp range from 0.105 at zero nose-out to a peak of 0.245 at 5° nose-out. The optimum tip speed ratio was between 2.1 and 2.5 depending on the nose-out.

Summary

The modified multiple streamtube model described at some length has predicted values of peak performance of 0.105 with zero nose out and 0.225 with $3\frac{1}{2}^{\circ}$ nose out. The field tests with similar geometries produced Cp's of 0.09 and 0.165. Drag from the crossarm will reduce the predicted output and although accurate assessment of the drag has not been possible calculations show that a drop in Cp of 0.03 at $\beta = 2.5$ is likely. Practical and Theoretical results are combined in Fig. 6.

The field tests showed much better performance at low tip speeds than predicted by the model and the optimum tip speed ratio was significantly less than predicted. The turbine started readily with both geometries.

Acknowledgments

The authors wish to thank the members of the Engineering Department who assisted in the field trials, especially Eric Preston who additionally was largely responsible for the construction of the turbine. The members of the Meteorology Department who provided both anemometers, data logging equipment and considerable advice. Finally we acknowledge SRC for Grant GRA/49423 which made the work possible.

References

1. Sharpe D.J. — "A theoretical and experimental study of the Darriers VAWT" Kingston Polytechnic Research Report 1977.

2. Strickland J.H. — "A performance prediction model for the Darriers turbine" Proc. Cambridge Wind Energy Systems Symposium 1976.

3. Walters R.E. et al. — "Innovative Straight Bladed Vertical Axis Wind Turbine" U.S. D.O.E. Wind Energy Innovative Systems Conference Colorado Springs. May 1979.

4. Stacey G. and Musgrove P.J. — "The Rutherford Six Metre Variable Geometry VAWT" 2nd BWEA Wind Energy Workshop. Cranfield 1980.

5. Abbott I.H. and Doenhoff A.E. — "Theory of Wing Sections" Dover Publications Inc. New York 1959.

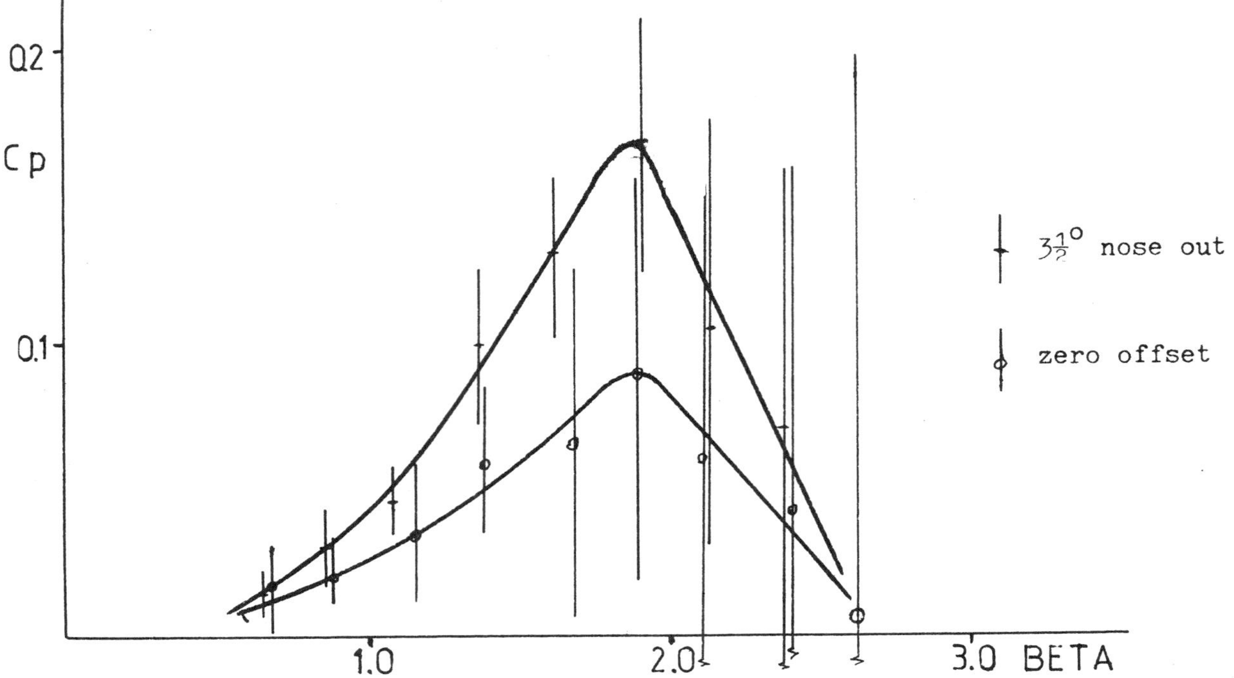

Fig. 1 Field Test Results

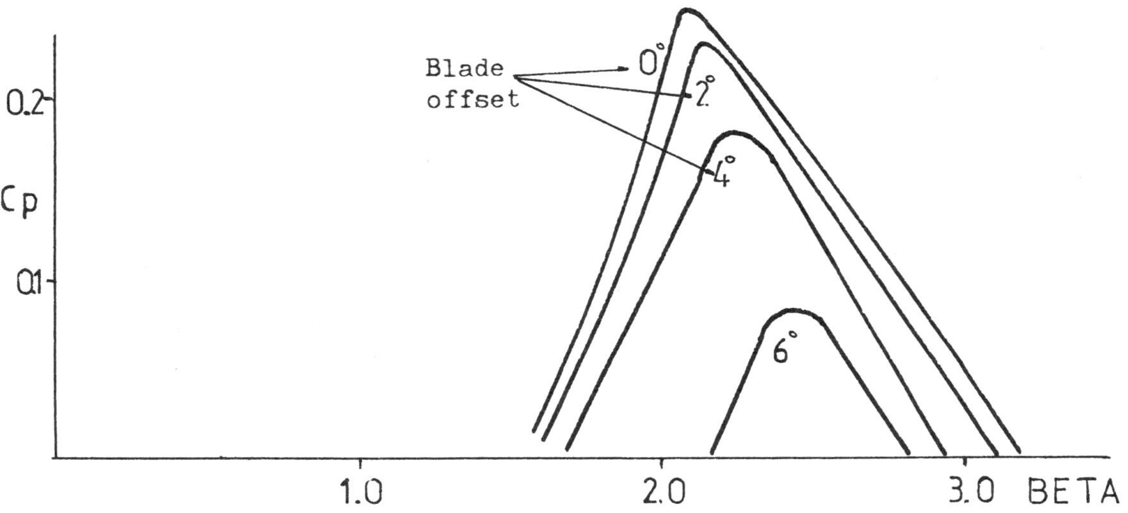

Fig. 2 Multiple streamtube predicted performance without virtual aerofoil effects.

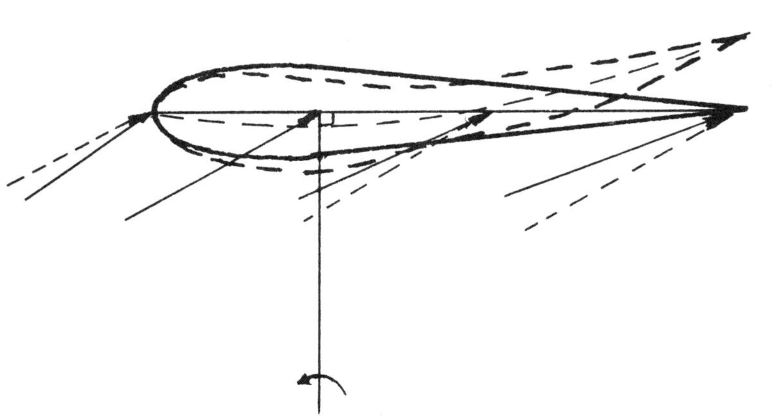

Fig. 3 Schematic aerofoil transformation

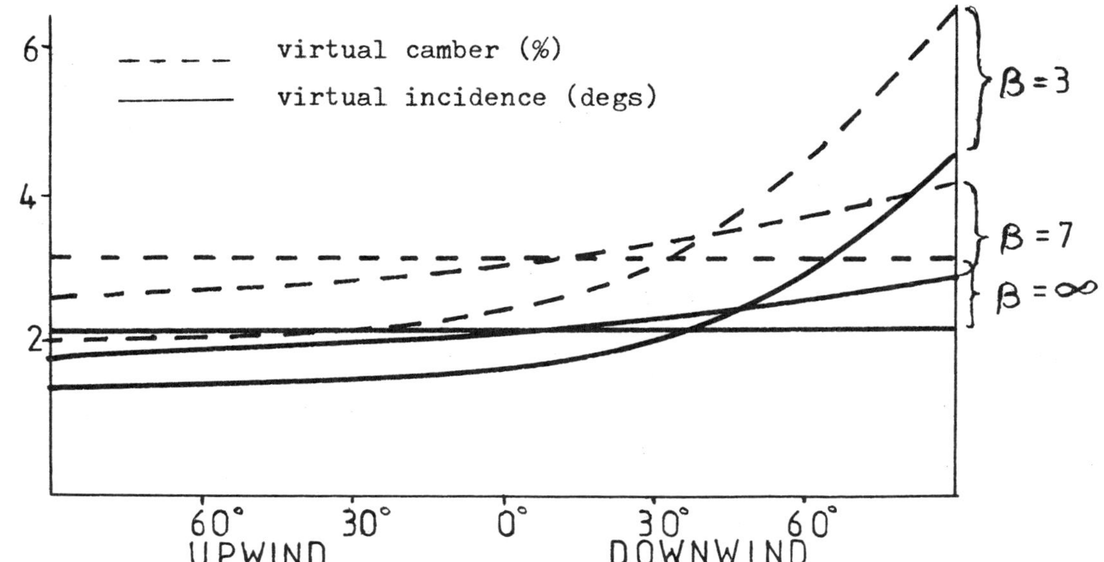

Fig. 4 Virtual incidence and camber at various tip speed ratios

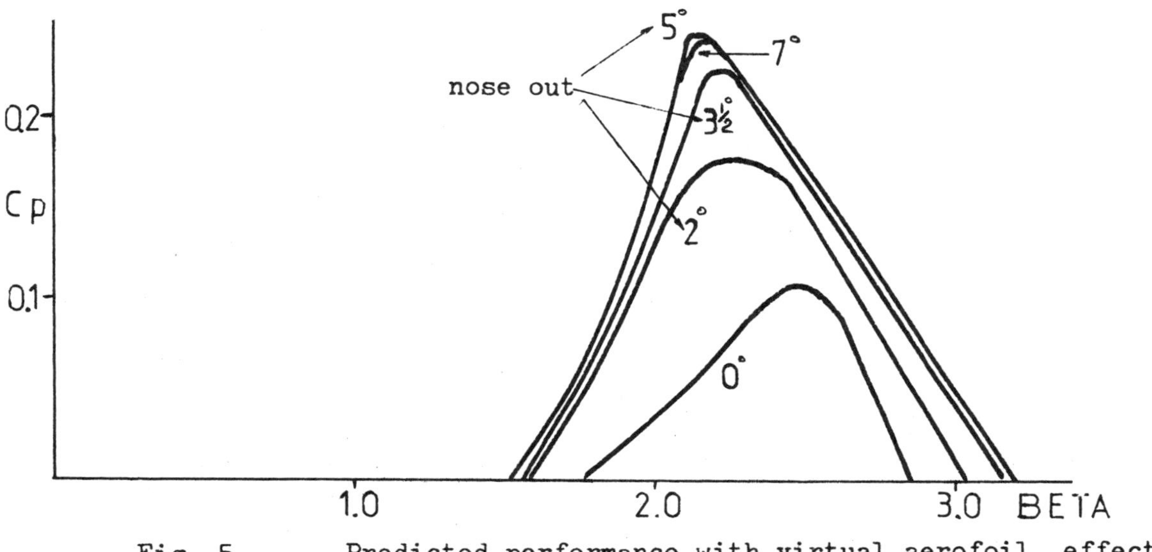

Fig. 5 Predicted performance with virtual aerofoil effects

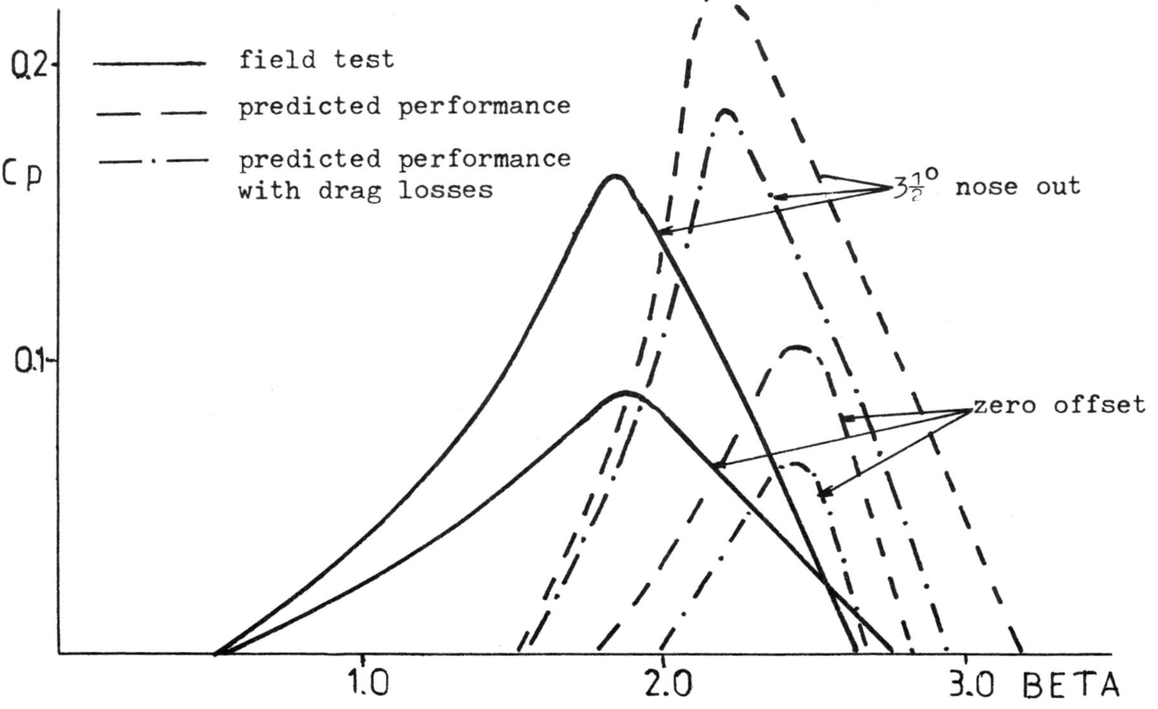

Fig. 6 Comparison of field test measurements with predicted performance.

WIND TURBINE WAKE STUDIES

B.R. Clayton and P. Filby
Department of Mechanical Engineering
University College London

Abstract

This paper provides details of the work carried out at UCL on the design and operation of an open-jet wind tunnel with a 1-metre square nozzle outlet and thus provides an update of the paper presented at the BWEA 80 meeting. Rotors of 0.5-metre diameter have been built and include models of larger scale horizontal and vertical-axis wind turbines undergoing tests in the field. Initial results are given of the tunnel calibration and also of the wake velocity distribution and momentum defect obtained from tests on a horizontal axis rotor on loan from the TNO, Netherlands. It is concluded that the facility, designed specifically for the purpose of model turbine testing, offers the versatility and accuracy required for detailed work on rotors and wakes and should therefore prove to be a valuable tool for the prediction of full-scale turbine and turbine array outputs.

Introduction

As stated in Ref (1) the two main problems concerning the prediction of full-scale power outputs from wind turbines lie in the difficulty of scaling from small models (of either single units, or even more severe, from large arrays) and the analysis of wake interference in large arrays, also subjected to 'scale effects'. It is imperative that predictions within ±2% over a year are achieved for the total energy generation (MWh) since very large total values are likely to be achieved with arrays having an installed total power of 2GW or so. Clearly, the problem of accuracy does not arise from fuel costs but from the capital, installation and maintenance costs of the turbines and associated equipment. Any shortfall in power, requiring added units, must be avoided otherwise heavy financial penalties accrue.

Perhaps the severest problem arises with the prediction of array performances owing to the very small size of the individual 'turbines' and the relatively low 'wind' speed used in wind tunnels (2,3,4). In fact, air speeds in the necessarily large wind tunnels are usually no greater than the design wind speed at full scale. So that a reasonably comprehensive array can be examined, a geometric scale as small as 1:1500 must be adopted and so Reynolds numbers of the full scale and model are also in this same ratio. Previous work with arrays has suggested that turbulence levels both in the wind reaching and aloft the array have an important bearing on the flow within the array. The latter consists of a continuous momentum exchange between the wake flow behind each 'turbine', the boundary layer flow near to the ground and the main-stream flow above. Just how these flows are affected by Reynolds number is not yet clear. Furthermore, owing to the small size of each component in the array, typically 70 mm diameter, geometric similarity between model and full scale turbines must be abandoned (other than to ensure that spacings between elements of the arrays are scaled in the same ratio as rotor diameters). Consequently, simulators in the form of circular screens of uniform mesh size, determined on the basis of the estimated momentum defect of an individual rotor, are used along with a downstream diffuser and placed perpendicular to the upstream approach flow. It appears that so far oblique flows and simulators at different heights in the array have not been examined. Some effort has been made to improve the simulator by placing a small anemometer rotor downstream from the screen. Whether this introduction of a rotary component to the flow along with, possibly, the generation of tip vortices provides improved accuracy is yet to be proven. As shown in (3) there is a rapid fall-off in power per row in the array with distance downstream so that by the time the fifth row, say, has been reached the power output from that row may be only 70 or even 50 per cent (depending on rotor spacing and orientation) of that from the first row. For a given energy flux in the wind an optimum to be aimed for is the maximum rate of extraction for a minimum size of the array. This may well call for a mix of types, tower heights and possibly a number of rotors of the same type but with blade geometries to suit optimum conditions within the array.

It was with the preceding background and development from it that led to the present programme consisting of the development of a rig to investigate, in detail, the flow in and around wind-turbine rotors and the characteristics of the downstream wake. Reynolds numbers up to the scale ratio times the full-scale Reynolds number can be achieved. Owing to the presence of turbulence and shear in the atmospheric boundary layer and the intense turbulent mixing within an array it may be that this disparity of Reynolds numbers may not be severe. However, it should be possible to assess the viability or otherwise of simulators in array models and thence to provide corrections in terms either of an empirical factor applied to the end result or to specify an improved simulator geometry. Further discussion of these matters may be found in Ref (1).

Experimental Rig

The dimensions of the rig are shown in Fig 1 and a view of the rig now constructed with tests underway is shown in Fig 2. As can be seen, the wind flowing on to the model rotor issues from a 1-m square nozzle with the rotor being placed 0.5 m downstream from the nozzle outlet plane. The purpose of this form of test facility is to allow a modest size of construction and yet to eliminate as far as possible the effects of blockage. That is, by not surrounding the rotor with solid walls, as in an enclosed wind-tunnel test section, it is expected that streamlines approaching the rotor may expand naturally on passing around the rotor without any forced contraction (and consequently increased velocities around the rotor tips). Two problems now arise. First, a 'blowing' source is required (in contrast to a 'suction' source from downstream fans in a straight-through wind tunnel) and this means that swirling turbulent flow from the fan source must be accommodated. Second, the jet issuing from the nozzle and subsequently the wake from the rotor mixes with the ground boundary layer and the still air in the laboratory and so a continuous wind for large distances downstream cannot be sustained.

A combination of screens and honeycomb straighteners is incorporated at different stations along the rig. The fan is a constant speed (24 rev/s), axial flow form with a continuously variable pitch angle. The latter can be adjusted by connecting a compressed air line across a pneumatic actuator so that the pitch angle can be altered with the fan running. Although not yet installed, automatic and continuous adjustment of the pitch angle can be effected by the installation of a servo feedback control using a comparison between the actual velocity required at the turbine and that currently obtained to provide the input to the feedback loop. Owing to the low pressure losses of this system the fan is working only slightly off its free-inlet-and-delivery conditon. Consequently, swirl downstream from the fan is small and can be easily removed by 100 mm of conventional impregnated paper honeycomb. Furthermore, wakes from the fan blades are also small, owing to small incidence angles at the fan blades, over the speed range called for at the jet. Far higher velocities than 10 m/s (the original upper limit) in the jet can be obtained with the fan still well off stall conditions and it is estimated that 20 m/s is possible. After initial problems with the first fan installed the second fan installed is now working satisfactorily.

Downstream from the fan the remainder of the construction is in marine plywood which retains dimensional tolerances over a wide range of ambient temperature and humidity conditions. Reinforcement with longitudinal wooden stiffeners and wooden blocks at corners has provided acceptable stiffness and absence of vibrations and 'drumming'. The diffuser transforms the circular cross-section of the fan duct to a square section of 2 m side length. This represents an increase in area of 4 to 1 over a length of 2.35 m. Wires stretched across the maximum area of the diffuser carried a series of tufts so that the flow could be viewed upstream from the nozzle. It was evident that the screens and honeycomb in the fan duct were successful in eliminating swirl and excess turbulence. The wake downstream from the fan motor was also successfully dissipated. However, it was noticed that some unsteadiness and slow flow occurred close to one side wall of the diffuser - possibly the result of non-uniform conditions at entry to the fan owing to the irregular geometry of the laboratory. Consequently, a graded screen was placed in position between the end of the diffuser and the entry to the contraction leading to the nozzle. This addition certainly improved the distribution of the flow. By the same token it is believed that the generation of shear flows to model the atmospheric boundary layer should not present too much difficulty. In this case, however, it is likely that screens in the contraction of the nozzle are likely to prove more useful.

The exit plane of the nozzle may be considered purely as that forming a jet in an infinite expanse of fluid or, alternatively as the exit plane for a jet exhausting into a semi-infinite medium from a location of small area situated in an infinite plane wall. In the first instance air will be drawn into the jet by the entrainment of air both surrounding and upstream of the contraction. Under th circumstances a considerable fall-off in velocity towards the edges of the nozzle observed after only a short distance downstream from the exit plane. It could well be that reference tests in uniform approach flows may not be reliable owing to the diameter of the rotor exceeding the span of uniform flow. As shown in Fig 1, the rig allows the blanking-off of a square of 2 m side surrounding the nozzle exit plane so that an approximation to the jet-in-wall case can be obtained. The effects of these two cases will be discussed in the following section.

A further point requiring investigation concerns the value or otherwise of turbulence stimulators at the periphery of the nozzle. The idea here is to encourage intense mixing at the edge of the jet with the stationary air in the laboratory and to thus preserve the uniformity and reduce the rate of decay of the jet with downstream distance. As we shall see, little benefit was obtained on the present rig; indeed, an undesirable velocity distribution resulted which penetrated well into the jet flow.

The eventual aim with the test rig is to mount the fan-diffuser-nozzle arrangement on rails so that the whole assembly may be moved longitudinally according to the length of groundboard required for wake traverses. The groundboard itself is in several sections of reinforced marine plywood so that the extent of the traverses may be varied. The rotor under test is mounted on a specially reinforced groundboard which can cater for horizontal and vertical-axis rotors. Each vertical-axis rotor has a central shaft supported at the top by a framework housing a bearing. The lower part of the shaft and its housing passes through the groundboard and is connected to a calibrated servo motor tacho-generator in the manner of many full-scale rotors at present under test. The arrangement is the same for both troposkien and straight-bladed Darrieus rotors.

Horizontal-axis rotors are mounted on an appropriate form of tower extending up from the ground board and on top of which is carried a similar power absorption device. In both cases the arrangement follows that already established by the Dutch workers who used a calibrated I.T.T. Dunkermotor Motor - tacho-generator with a speed control operating off the tacho-generator output. Power was measured from the electrical output plus a correction for motor losses.

A traversing carriage, supported on a gantry, is at present manually operated. Precision rails are fixed to the top of the gantry, all of which is well clear of the air flow, and the carriage runs freely on the rails. The gantry itself will ride on the same rails as those which support the remainder of the rig so that easy transverse and longitudinal travel of the traversing instrumentation is ensured. Vertical movement of the instrumentation is via a slide adjacent to a fine scale. Although such an arrangement is simple, reliable and robust it does mean that a complete cross-section traverse of the wake of a turbine rotor is time consuming. Experiments are now well in hand on the perfection of automatic traversing on the basis of a computer programmed input schedule with eventual output in the form of a graph plotter and printout of digital data.

So far measurements of velocity and pressure have been taken with a pitot-static tube and wall tappings connected to a micromanometer. A measure of centre line turbulence level using a hot-wire anemometer has also been undertaken in three mutually perpendicular directions.

A model has been constructed of the Swansea University 5-m machine with the blades and hub being machined from solid aluminium alloy on a numerically controlled milling machine. This departmental facility is immensely useful in constructing complex shapes from a digital input and items constructed in this way range from 1.25-m long ship models (from wood), joints for artificial limbs and robots, aerofoil sections with and without twist and so on, down to aerofoils from steel having a span of 50 mm, a chord length of 25 mm and a maximum thickness of 2.5 mm. The blades and cross-arms of a straight-bladed Darrieus model have also been constructed quite simply. (A 2-m diameter straight-bladed Darrieus with wooden blades is nearing completion before field tests commence shortly.) So far, tests have been restricted to those on the TNO horizontal axis rotor which has been loaned to the authors. The rotor, 0.36 m diameter, consists of two steel, untwisted blades fixed to a boss extending from a small calibrated generator and speed controller. This assembly is mounted atop a steel, triangular framework, which in the present study is bolted in turn to the groundboard. A similar construction will be used for the Swansea U. rotor model unless full-scale wake measurements indicate that the tower and supports plays an important role in both the near and far wake development eg by the generation of a persistent trailing vortex.

To supplement the experimental programme on model rotors a series of experiments is planned, with student assistance, to study in detail the wake formation and propagation from screens with and without diffusers and rotors, of up to 0.5-m diameter in order to provide data on Reynolds number effects and the appropriateness of such devices as array simulators. Oblique and shear flows will also be included.

Calibration of the Open-Jet Rig

The justification for the present form of the rig, described in the previous section, was based on a series of tests. Figure 3 shows the difference between the vertical velocity distribution with the jet-in-space (a) and the jet-in-wall (b) configuration obtained by blanking-off the 'windows' in the framework at the

For these tests run at 9.5 m/s vortex generators, in the form of rectangular metal strips, were fixed at regular locations round the periphery of the nozzle and extended some 50 mm into the jet. It is seen that at 0.5 m (nominally one rotor diameter) from the nozzle outlet plane the effects of entrainment were still strong even with the 'windows' closed. When the groundboard was installed, of length 0.5 m, matters were significantly improved, as shown in **Fig 3(c)**, with the same improvement evident when a 'roof' of equal size was incorporated at the top of the nozzle. Clearly, this arrangement could not be repeated for the sides otherwise the original nozzle arrangement (extended) would have resulted. However, this 'ground and roof' did allow a uniform velocity distribution over the test section for the basic flow of rotors to be obtained.

Velocity traverses were then completed across the whole flow area in the plane that the rotors would occupy. Of particular interest were the velocity traverses, again at the centre line velocity of 9.5 m/s, at 0.5 m downstream from the sides of the nozzle with the vortex generators in place and then removed. As **Fig.4** shows be seen there is a considerable improvement in the uniformity of the flow and so the vortex generators have not since been used. Owing to the more even distribution of the flow the velocity further into the jet decreased by about 4 per cent and became even more uniform. In Fig 5 are shown corresponding horizontal velocity traverses. Figure 6 shows an isometric plot of the complete velocity profile across the plane corresponding to model rotors and is that adopted for the standard 'uniform flow condition' applied to the different models. Also shown on Fig 6 are the locations of the planes corresponding to those at which the traverses depicted in Figs 4 and 5 were taken.

It is known that turbulence levels in the main air flow have a significant effect on the geometry of the wake downstream from the rotor. There is, as yet, no definitive correlation between turbulence levels and e.g. momentum defect, rotor power, shear flow variation and so on. Nor do the present authors accept that only longitudinal velocity fluctuations need be considered. Turbulence is not isotropic in the present context and this is so for the present rig. Figure 7 shows measured centre line turbulence levels of the free jet, i.e. with the rotor and tower removed. It may be noted that the turbulence levels, deduced from hot-wire anemometer readings, are high (by a factor of 2) compared with many enclosed wind tunnels. Nevertheless, these levels are likely to be more realistic for full-scale work. The effects of these high turbulence levels can be seen in the following section and as a result the most immediate efforts will concentrate on the use of screens in the settling chamber to reduce the overall levels. At a plane corresponding to 0.6 m downstream from the rotor plane transverse and vertical turbulence levels were measured and found to be rather higher than the corresponding longitudinal level. The latter first decreased with distance downstream from the rotor as momentum exchanges took place in the inner portion of the jet. At greater distances than 2 m an increase was observed as turbulence from the outer edge of the jet, resulting from air entrainment, was convected inwards. It is unlikely that traversing beyond 5 m downstream from the rotor plane would yield useful results. To date no turbulence levels have been taken with a rotor in position and turning.

An important aspect of the future experimental programme concerns a detailed investigation of the wake development when the turbine rotor is subjected to an entry shear flow, such as that experienced by turbines immersed in the atmospheric boundary layer. Some preliminary attempts to generate a shear flow across the model rotor test plane have been completed with the results shown in Fig 8. The shear flows A and B were obtained with different arrangements of slats at the nozzle exit plane. The accompanying turbulence level was high but not unrepresentative of wind turbulence at full scale. Nevertheless, the next step in this part of the modelling is to use a combination of screens (curved at the nozzle in order to present a non-uniform solidity to the air flow) and vortex generators to simulate a range of turbulence levels. As can be seen in Fig 8 the experimental values encompass the 'one-seventh' law for boundary layers over flat ground and the 'one-tenth' law for boundary layers over clam water. Incidentally, these latter curves were developed using a 600 m standard mixing length with a 10 m/s wind at hub height. Some difficulty in the generalisation of results may be anticipated for shear flow studies. Because the variation of incident velocity with height is non-linear the performance of a given turbine will depend on both hub height and rotor diameter which together are needed to specify the radial velocity variation. Although the velocity distribution tends to linearity as altitude increases, and therefore allows simpler scaling laws for very large machines, the same is not true of smaller rotors. A point to be examined, therefore, concerns the accuracy of scaling rotors in the 10 m diameter range to predict performances of rotors in the 70 - 100 m diameter range. The response of different types of rotor to shear flows is also, of course, of primary concern.

Tests with the TNO Model Rotor

Figure 2 shows the rotor in the position for which all test results have been obtained so far. Figure 9 shows the power coefficient $C_p = P/\frac{1}{2}\rho AV^3$ as a function of the blade-speed ratio $\lambda = u/V$, where P is the shaft power which includes a correction for motor losses, ρ is the density of air, A is the area of the rotor disc, u is the rotor tip speed and V is the upstream approach velocity in Ref (5) and the velocity at the rotor plane in the absence of the rotor in the present work. The data were determined for the same operating points as those previously deduced at the TNO, Netherlands (2), namely for values of λ given by a = 5.0, b = 6.6 and c = 8.5. It is clear that close correlation exists although it must be emphasised that data were taken at only three values of λ and at the same approach velocity of 9.5 m/s. Turbulence levels appear to have only a minor effect on C_p.

Velocity traverses were taken at 1.67 and 9.2 rotor diameters (0.601 m and 3.312 m) downstream from the rotor for the values of λ corresponding to a, b and c on Fig 9 in order to make a direct comparison with the TNO data. The first results taken were for a vertical centre line traverse at a distance of 1.67 diameters downstream from the rotor plane as shown in Fig 10. The extent of the traverse was ±0.65 m about the horizontal centre line through the nozzle although for the sake of clarity only the range ±0.5 m is shown on Fig 10 and subsequent figures. At the two higher values of λ a distinctly flat portion of the velocity profile was observed to extend over a plane having a diameter about one-third that of the rotor, i.e. about 0.12 m. This is no doubt due, in part, to the presence of an extending wake downstream from the hub and motor. No vertical traverse was published in (5) but a horizontal traverse at the same location is given and this displays a somewhat strange behaviour for λ = 5.0. The TNO results are shown in Fig 11 along with the present data and the odd local minimum in the TNO data is certainly not displayed by the UCL open-jet rig results. It was thought that maybe a persistently shed trailing vortex was responsible. However, detailed traverses in that region at different distances from the rotor revealed no such behaviour and so we feel that this is a spurious result from (5) which is not revealed in the present experiments. Neither was there any evidence from the vertical traverses of shed vortices. Nevertheless, some small differences were evident between the vertical and horizontal traverses but this could be the result of the downstream wake shed by the tower since this was not axisymmetric.

There is evidently, a significant difference between the velocity profiles from Ref (5) and those obtained by the present authors and yet both result from carefully conducted traverses. There could be several reasons for this. In Ref (5) the height of the tunnel was only 1.2 m i.e. less than four times the rotor diameter. It is therefore possible that the wake shed by the rotor was artificially contracted by blockage effects which tended to increase the mainstream flow away from the wall boundary layers and thus exaggerate the velocity defect at the centre of the wake. Alternatively, the higher turbulence levels prevailing in the open-jet rig may tend to 'even out' the wake and certainly the effect of turbulence on the velocity defect at the centre-line indicates that this effect is present.

At the downstream plane 9.2 diameters behind the rotor the effects of the wake appear to be negligible in that no velocity deficit in an identifiable wake can be detected as shown in Fig 12. Notice that the centre-line velocity has decreased (for the free jet) only from 9.5 m/s to 9.2 m/s and that the velocity at the edges of the traverse shown are little different from those in e.g. Fig 10. Only results for a vertical traverse are available but we may conclude that the free jet is still acting as though it were a sustained wind without the confining boundaries of a wind tunnel. The reduction in momentum as a result of energy extraction by the rotor appears as an even velocity defect across the flow area.

The velocity defect along the horizontal centre line measured at distances downstream from the rotor is depicted in Fig 13 in a way similar to that shown in Refs (3,5). We again confirm that the shape of the curve on logarithmic scales could be considered within the range -1.25 to -1.4 but the curves are displaced to the left of those from Ref (5). This is almost certainly the result of greater mainstream turbulence levels and more intense mixing of the wake flow with the outer flow.

A final note of caution should be sounded in respect of the detailed trends of the present results. A rather special kind of rotor was used and this is not really representative of real designs. Thus, although the value of the open-jet rig is well-founded it would perhaps be injudicious to read too many general trends into the results until far more data are obtained and the results comprehensively processed.

Conclusions

The open-jet rig described in the previous sections is believed to represent an accurate and versatile arrangement as a purpose-built facility for testing model turbines up to a diameter of 0.5 m. The immediate use of the equipment is related to a wake studies programme on several types of rotor and to relate the results to scale effect on wakes and the relevance of simple simulators for array tests. However, it is considered that the rig has great versatility and it is intended to extend its use, through increased automation, to cover detailed rotor studies, unsteady flows, structural analysis and material tests under hazardous conditions. Results are presented of the wake behaviour of a previously tested rotor in order to check the data and to provide support for the open-jet facility. This has been done and, furthermore, previous inexplicable behaviour has not been verified and it is suggested that the earlier data could be spurious.

Acknowledgements

The authors wish to thank Dr. P. Builtjes for arranging the loan of the TNO rotor. The work reported here is supported by an SRC grant.

References

1. CLAYTON, B.R. and FILBY, P. BWEA Wind Energy Workshop, Cranfield, 1980.

2. BUILTJES, P. Proc. Second International Symposium on Wind Energy Systems, Amsterdam, PaperB5, 1978.

3. BUILTJES, P. and MILBORROW, D.J. CERL Rep. No. RD/L/N72/80, 1980.

4. ALFREDSSON, P.H. and DAHLBERG, J.A. Aeronaut. Res. Inst. Sweden, Tech. Note AU-1499, 1979.

5. VERMEULEN, P. TNO Apeldoorn, Netherlands, Rep. 78-09674, 1978.

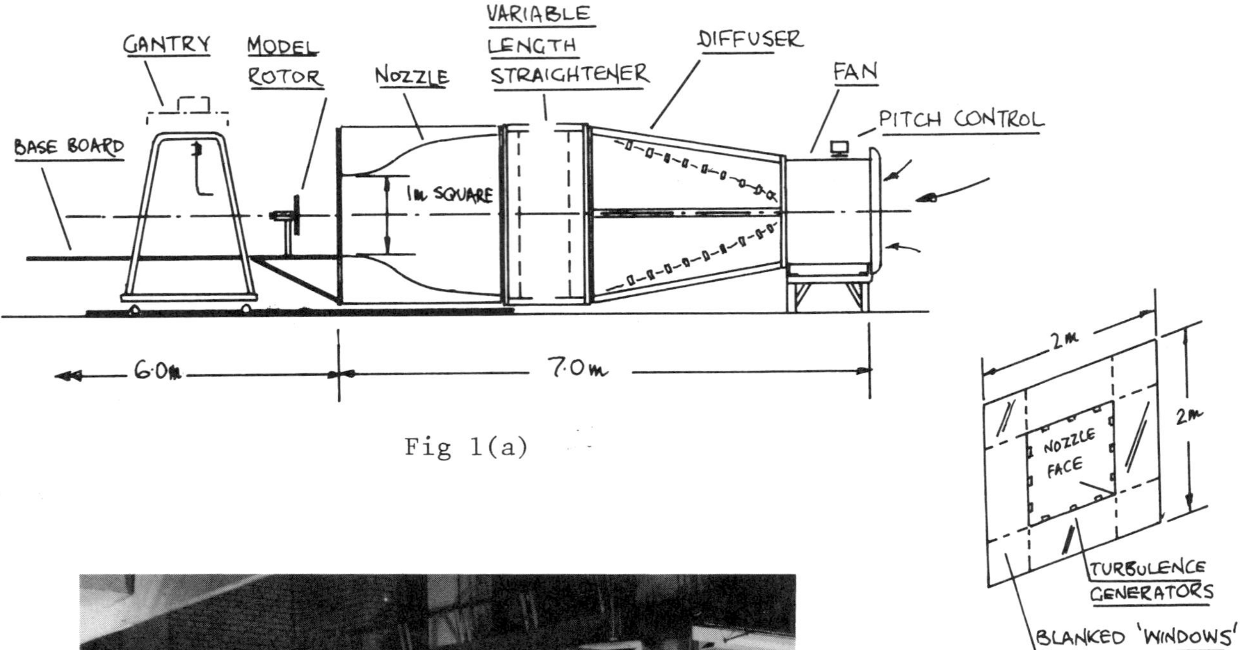

Fig 1(a)

Fig 1(b)

Fig 2(a)

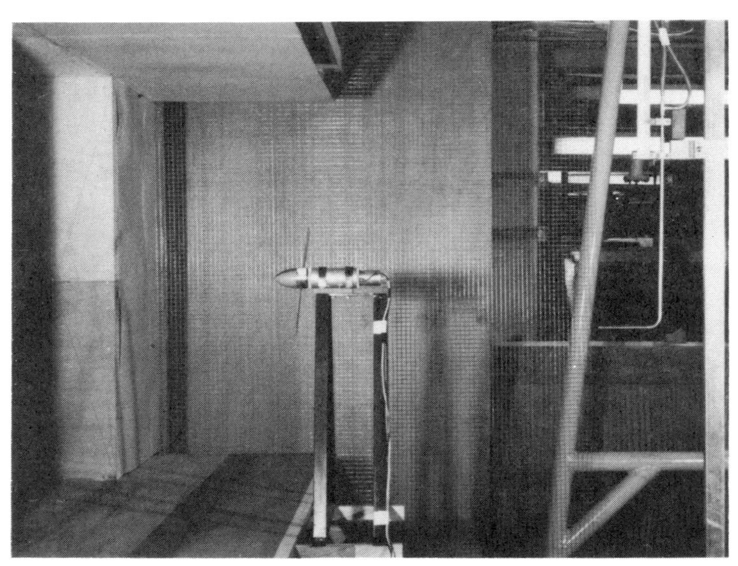

Fig 2(b)

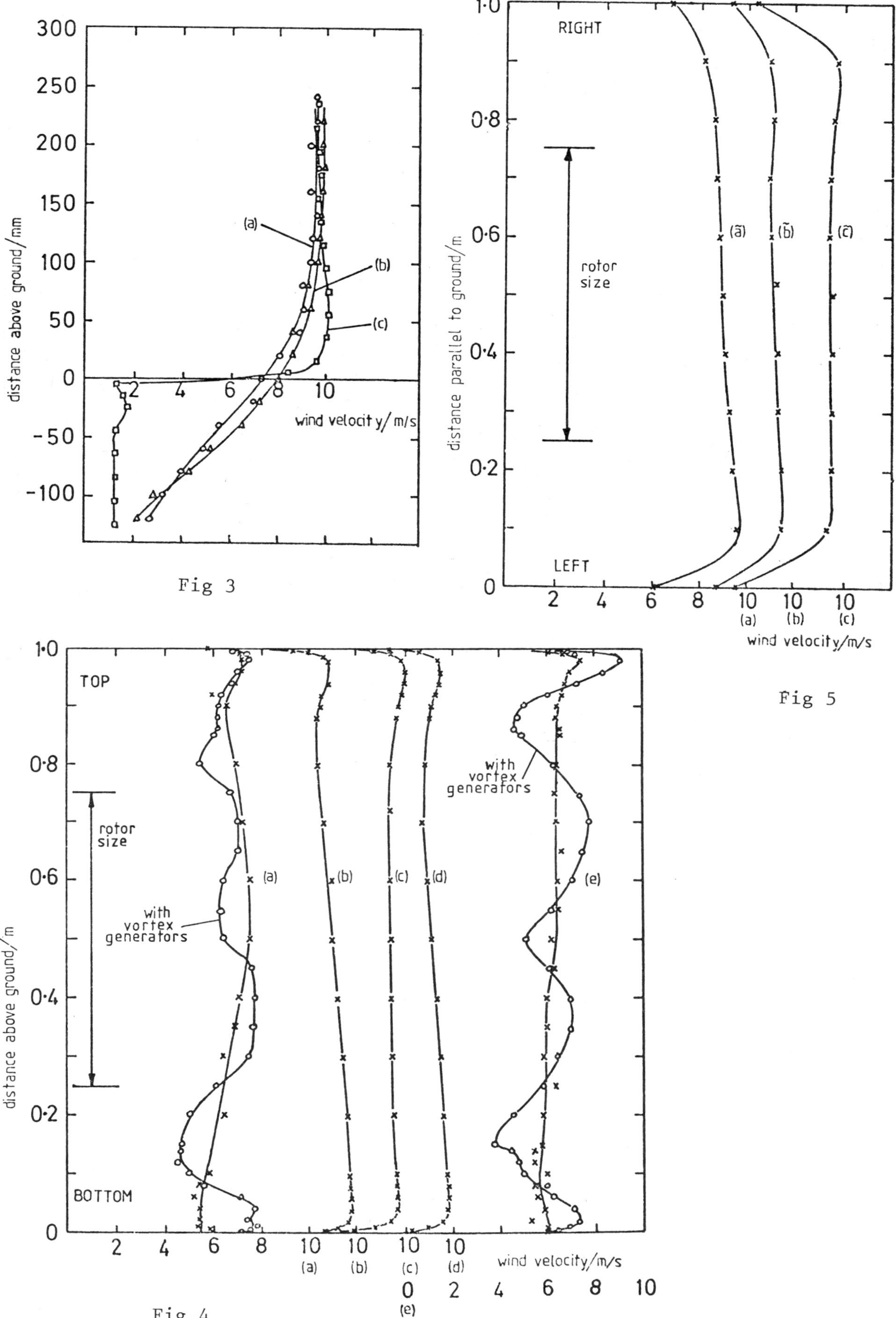

Fig 3

Fig 5

Fig 4

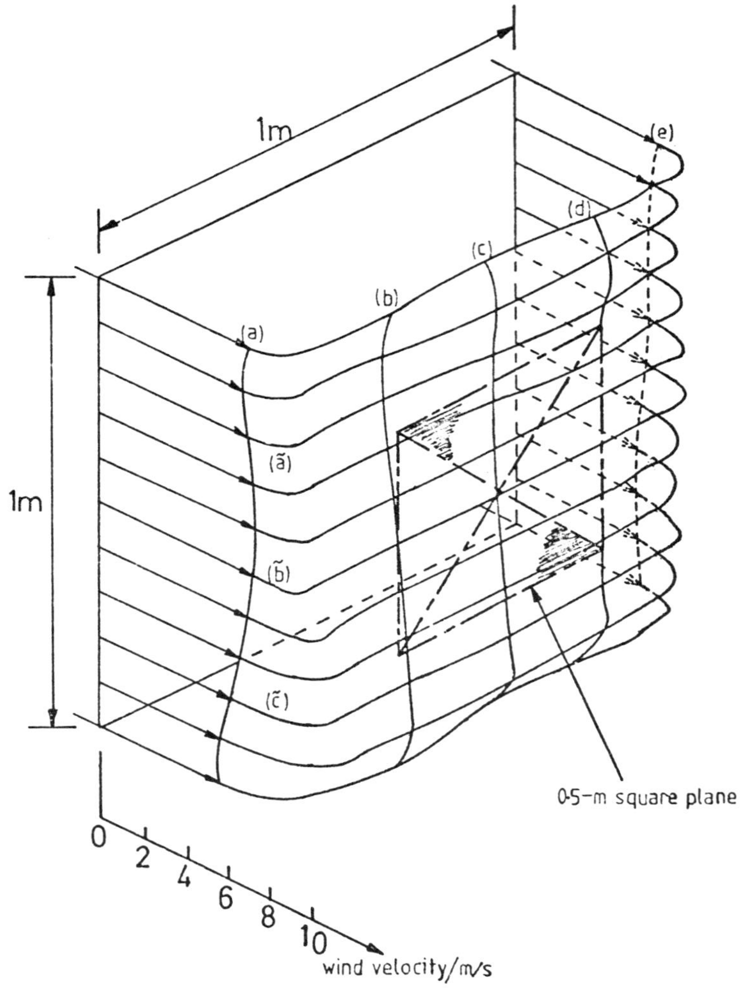

1m
1m
(a)
(b)
(c)
(d)
(e)
(ã)
(b̃)
(c̃)
0-5-m square plane
0
2
4
6
8
10
wind velocity/m/s

Fig 6

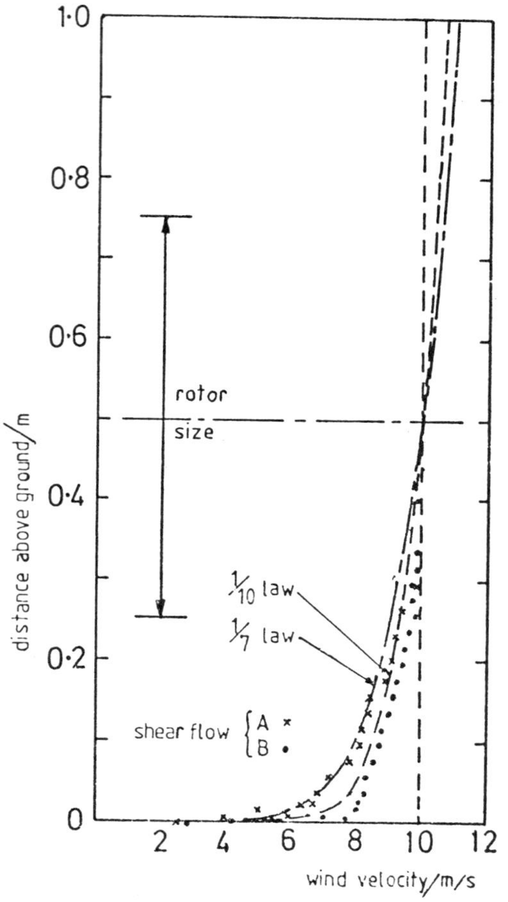

1·0
0·8
0·6
0·4
0·2
0
rotor
size
distance above ground/m
1/10 law
1/7 law
shear flow
A ×
B •
2
4
6
8
10
12
wind velocity/m/s

Fig 8

20
18
16
14
12
vertical
transverse
longitudinal
turbulence level/%
0
1
2
3
4
distance downstream from rotor/m

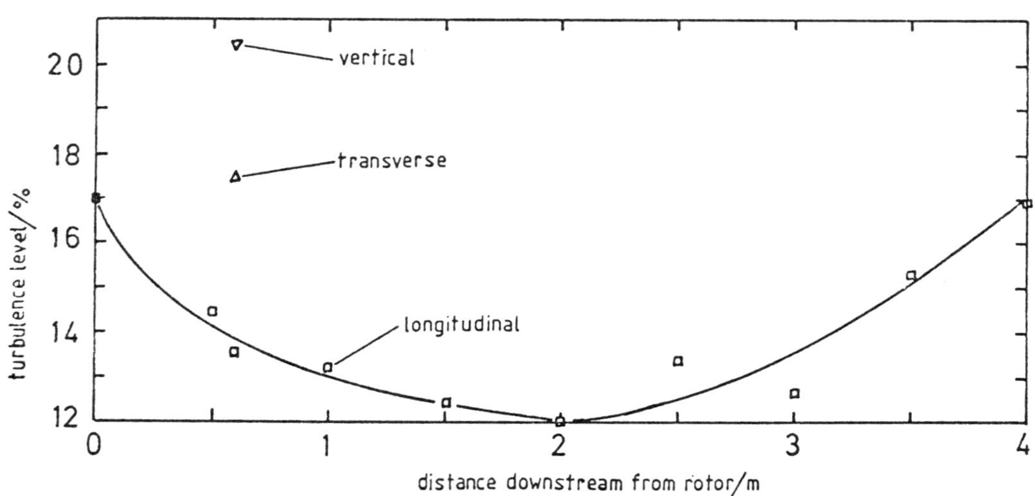

Fig 7

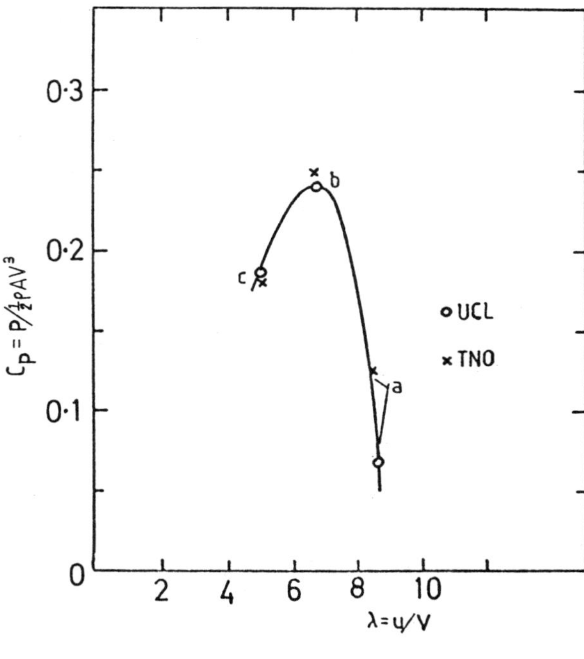

Fig 9

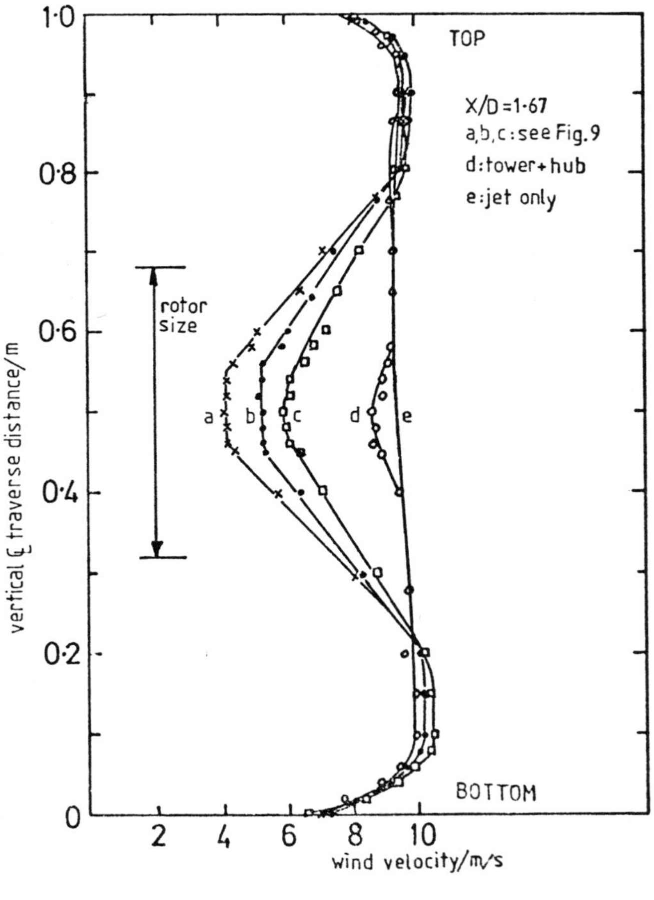

Fig 10

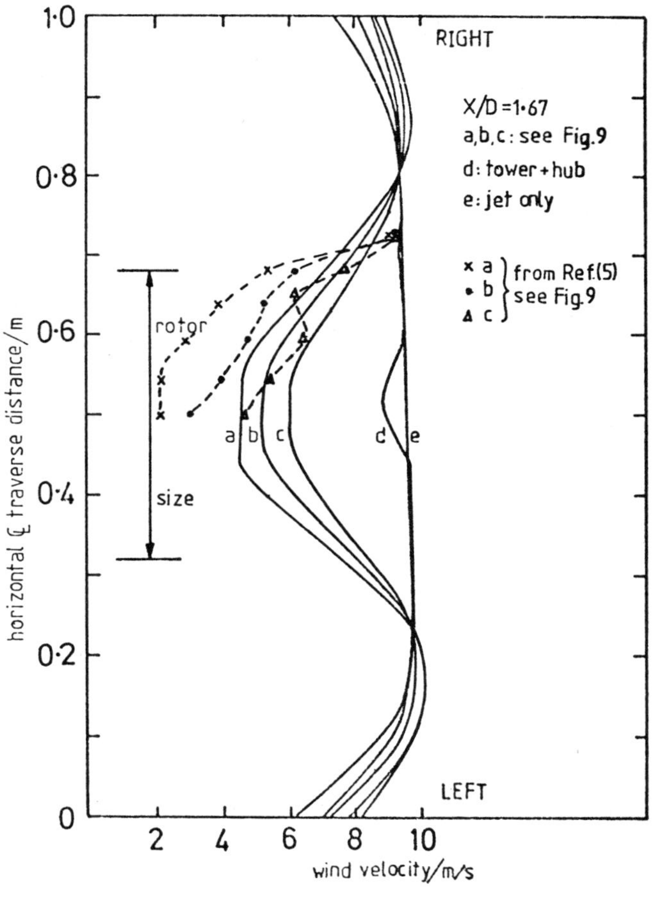

Fig 11

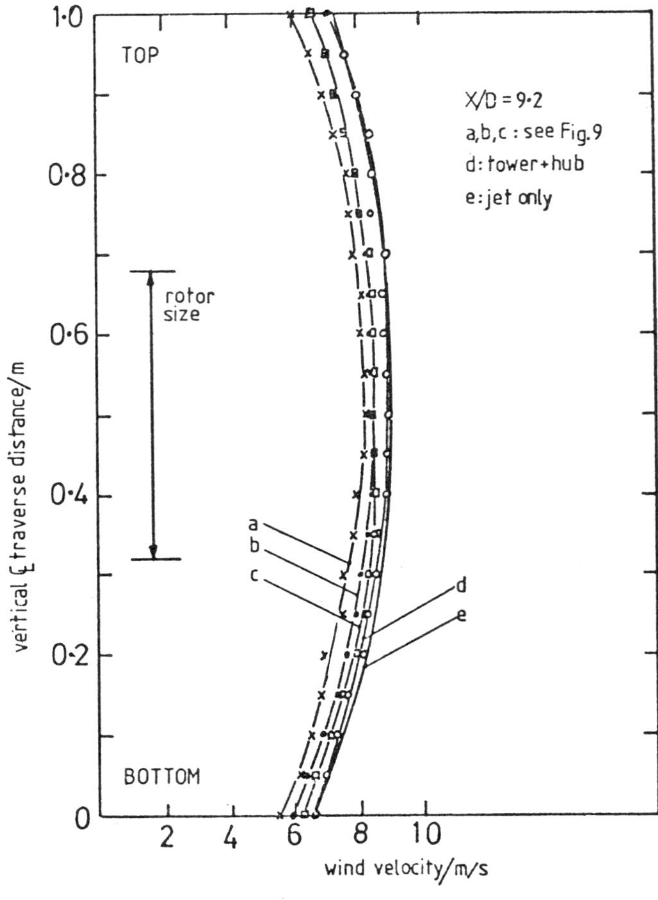
TOP
BOTTOM
X/D = 9·2
a,b,c : see Fig.9
d: tower+hub
e: jet only
rotor size
vertical ℄ traverse distance/m
wind velocity/m/s
a
b
c
d
e

Fig 12

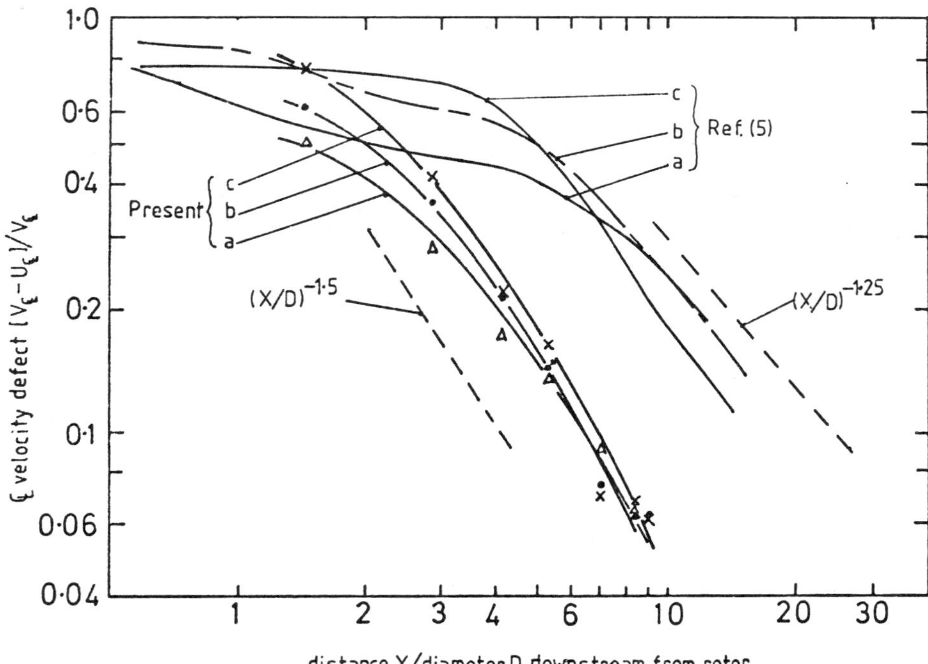
℄ velocity defect $[V_℄-U_℄]/V_℄$
distance X/diameter D downstream from rotor
Present
Ref. (5)
$(X/D)^{-1.5}$
$(X/D)^{-1.25}$

Fig 13

WAKE MEASUREMENTS IN CLUSTERS OF MODEL WIND TURBINES USING LASER DOPPLER ANEMOMETRY

J.N. Ross and J.F. Ainslie

CENTRAL ELECTRICITY RESEARCH LABORATORIES

LEATHERHEAD

Abstract

A series of wind tunnel experiments is being carried out to investigate the wake effects in clusters of wind turbines. Windspeed and turbulence levels have been measured both in single wakes and deep within the cluster, using a laser doppler anemometer.

Results are presented which can be compared with wake measurements behind full scale wind turbines. The measurements of air flow within the cluster will also enable some of the assumptions used in theoretical cluster models to be evaluated. Possible implications of the high turbulence levels measured are also noted.

INTRODUCTION

There has been a growing interest in examining the performance of wind turbines grouped together into clusters, to determine whether appreciable quantities of power can be generated within a compact land or sea area. The presence of wind turbine wakes within a cluster will reduce the power output from the cluster, and previous studies have shown that the effects can be large (Milborrow, 1980).

Wind tunnel modelling offers a quick and flexible experimental technique for investigating wake effects within clusters. Estimates of cluster performance have been made using model wind turbines; there is, however, a lack of data on the flow field within a cluster. In view of this, and the suitability of the laser doppler anemometer to the remote measurement of air speed and turbulence, a comprehensive set of flowfield measurements has been made using a cluster of wind turbine models in the low speed wind tunnel at CERL.

OBJECTIVES

The objectives of this study are:-

1. The provision of wake data for comparison with other wind tunnel experiments on simulators and rotors, and for comparison with full-scale wind turbine wake data.

2. To gain an improved understanding of wake effects within a cluster, to enable the assumptions used in theoretical techniques to be evaluated.

3. To quantify the increased turbulence levels within a cluster, and provide data on velocity profiles within a cluster, since these have relevance to wind turbine design and structural integrity.

LASER ANEMOMETER

Method

The laser anemometer used was of the 'Doppler difference' or 'real fringe' type. The optical arrangement is shown in figure 1. The output of the laser is split into two beams which cross in the region from which the velocity is to be sampled. The detector collects light scattered from small particles suspended in the air as they pass through the sample volume and it is the velocity of the scattering particles which is measured (see for example Abbiss, 1977). The light scattered by a particle passing through the sample volume is modulated at a frequency f given by:

$$f = 2\, u \sin(\tfrac{1}{2}\theta) / \lambda$$

where u is the velocity component in the plane of the two laser beams and perpendicular to the direction bisecting the angle between the two beams; λ is the wavelength of the laser light and θ is the angle between the beams.

By monitoring the light scattered back towards the laser, the laser, beam splitter and detector can be rigidly mounted to maintain alignment, and the whole assembly moved to traverse the sample volume. Using this 'backscatter' geometry and a HeNe laser of relatively low power (5 mW) satisfactory signals were obtained if the wind tunnel was 'seeded' with a small quantity of smoke to provide the scattering particles.

The size of the sample volume is determined by the diameter of the laser beams at the crossing point, and the angle between the beams. For the measurements reported here the sample volume had a diameter of about .34 mm and a maximum length of about 11 mm. The effective length of the sample volume will be rather shorter than this as the signal is weighted towards the centre of the sample, where the laser intensity and the number of fringes are greatest.

The signal from the detector is analysed with a photon correlator which produces the autocorrelation function of the intensity received by the detector. From the autocorrelation function the modulation frequency, and hence the velocity, is deduced using the curve fitting technique of Birch, Brown and Thomas (1975). This procedure represents the velocity distribution function P(u,v) by a truncated Gram-Charlier distribution,

$$P(u,v) = (1/2\pi\sigma_u\sigma_v)\exp\{-(u-\bar{u})^2/2\sigma_u^2 - v^2/2\sigma_v^2\} \times \Sigma A_k He_k\{(u-\bar{u})/\sigma_u\} \qquad \ldots \quad (1)$$

where v is the velocity component perpendicular to the plane of the laser beams. σ_u, σ_v, $\bar{u}$, $\bar{v}$ and A_k are constants, He_k is a Hermite Polynomial of order k. Note that if only the first term of the summation is retained, the distribution reduces to a Gaussian distribution. A model of the autocorrelation function can be derived analytically using (1) to represent the velocity distribution. This model function is fitted to the experimental autocorrelation function by a least squares method and the statistical properties of the velocity component are derived (mean, rms deviation, skewness and kurtosis). All the measurements reported here used the first three terms of the expansion.

ACCURACY

The two factors which determine the accuracy with which the velocity can be measured are the angle between the laser beams and the accuracy with which the frequency can be deduced from the correlation function. The angle between the laser beams is found from the separation of the beams at the beam splitter and the distance to the sample volume. For the measurements reported here the uncertainty in the angle between the beams is estimated to be ±1°/.

The accuracy with which the mean frequency is derived from the correlation function is much more difficult to assess because it depends on how well the assumed model for the distribution can fit the actual distribution. An attempt has been made to estimate the uncertainty by comparing different records made under the same flow conditions and then using different numbers of terms in the Gram-Charlier model of the distribution. The uncertainty in the mean increases as the turbulence intensity (standard deviation of fluctuations divided by the mean) increases and also increases if the distribution is not symmetric (ie. has a large skewness). For a reasonably symmetric distribution with a turbulence intensity in the range 0.1 - 0.3 it is estimated that the uncertainty in the mean is in the range 2-5%. The uncertainty in the turbulence intensity is probably about 10% of the turbulence intensity over the same range of conditions. The accuracy with which the sample volume was positioned was about ± 1 mm in height, ± 2 mm across the tunnel and ± 2 mm along the axis of the tunnel.

MODELLING TECHNIQUE

The simulation technique used was based on that used by Milborrow (Builtjes and Milborrow, 1980). The wind turbine models consisted of 72 mm diameter vane anemometers, with gauze attached to the upstream face of the anemometers to simulate the extraction of power from the airstream. The stems of the anemometers were sunk into the wind tunnel floor and the hub height was one model diameter for the tests reported here. The thrust coefficient of the simulators was increased in some of the experiments by attaching a second gauze to the rear face of each anemometer. The thrust coefficient of the single gauze was estimated from wake measurements as 0.66, that of the double gauze combination as 0.9.

A Counihan type boundary layer simulation technique was used in the wind tunnel (Counihan, 1972), the 1:500 scale boundary layer incident on the cluster being approximately a 1/9 power-law profile. The ambient longitudinal turbulence at hub height was ∿ 17% based on the hub height mean windspeed (14% based on the windspeed at the top of the boundary layer).

RESULTS

Single Wake

The velocity deficit $(1 - u/u_o)$ contours close behind the model (fig 2) show the vertical asymmetry of the wake caused by the support stem of the simulator. Fig 3 shows the downstream development of the wake. In Fig 4 the vertical wake profiles are compared with results from similar experiments at TNO, the Netherlands (Vermeulen, 1979), for which the thrust coefficient was 0.85. This shows that the CERL simulator support stem had a larger wake deficit than the support of the TNO simulator. Otherwise the agreement is good.

The centreline velocity deficit and turbulence properties of the wake are shown in Fig 5. The centreline velocity deficit decays at a similar rate to that which Vermeulen (1980) found to be characteristic of both model rotors and simulators. The centreline turbulence level is very high close to the rotor, but decays fairly quickly to ∿ 25%. The turbulent velocity shows a peak of ∿ 23% of u_o (the incident hub height mean velocity) between 2 and 4 diameters downstream. The data of vermeulen (1978) on a horizontal axis model rotor shows a similar peak between 4 and 10 diameters downstream, the position of the peak being dependent on thrust coefficient.

Measurements within the cluster

The cluster configuration used was the same as that used by Builtjes (1978). It consisted of 37 models arranged in a regular hexagonal array, with 7-diameter spacings.

Fig 6 shows the hub height velocity and longitudinal turbulence intensity incident on each row of the cluster, measured 2 diameters upstream of each row, with the wind blowing parallel to the columns of the array. The hub height velocity drops quickly near the front of the cluster, and the velocities incident on each row are similar for both values of thrust coefficient. The turbulence intensity rises rapidly to reach a steady value; for the high thrust coefficient this asymptotic value is a little higher than for the lower thrust coefficient, but the initial rise in turbulence is much steeper. The asymptotic turbulence levels are a little higher than those suggested by Milborrow (1980) on the basis of boundary layer theory.

It would seem that the higher thrust coefficient results in an increased turbulence level, and hence an increased rate of mixing. The result is that the velocity decay through the cluster is little changed by an increase in thrust coefficient.

The change in vertical profiles of velocity and turbulence within the cluster are shown in figs 7 and 8. These measurements were made 2 diameters upstream of the first, second and seventh rows. The velocity profile is dramatically changed by the presence of the cluster. The profile at the second row is dependent on the thrust coefficient, but by the seventh row there is little difference. The change in the turbulence profile of the flow is also large. For the high thrust coefficient models, the turbulence levels vary little after the second row, and all the turbulence profiles within the cluster have a characteristic shape with the peak intensity below hub height.

CONCLUSIONS

The laser doppler anemometer has proved to be an accurate and versatile instrument for making comprehensive measurements of windspeed and turbulence in wind tunnel studies of wind turbine wakes. The single wake data that has been collected will enable comparisons to be made with wake measurements behind full-scale wind turbines operating in the real turbulent boundary layer. The measurements of air flow within the cluster will enable some of the assumptions used in theoretical cluster models to be evaluated. Turbulence intensities above 25% have been measured in the flow incident on wind turbines deep in the cluster, and the effect of this on the structural and aerodynamic behaviour of wind turbines should be assessed.

ACKNOWLEDGEMENTS

This work was carried out at the Central Electricity Research Laboratories and is published by permission of the Central Electricity Generating Board.

REFERENCES

Abbiss, J.B., 1977, "Photon Correlation Velocimetry in Aerodynamics" in "Photon Correlation Spectroscopy and Velocimetry", ed. Cummins and Pike, Plenum Press.

Builtjes, P.J.H., 1978, The Interaction of Windmill Wakes, Proc. 2nd int. Symp. on Wind Energy Systems, Amsterdam, B.H.R.A. Cranfield.

Builtjes, P.J.H., and Milborrow, D.J., 1980, Modelling of Wind Turbine Arrays, Proc. 3rd. Int. Symp. on Wind Energy Systems, Copenhagen, B.H.R.A. Cranfield.

Birch, A.D., Brown, D.R. and Thomas, J.R., 1975, Photon Correlation Spectroscopy and its application to the measurement of turbulence parameters in fluid flows, J. Phys D : Appl. Phys., 8 p.438-447.

Counihan, J., 1972, The structure and wind tunnel simulation of rural and adiabatic boundary layers, Symp. on External Flows, Univ. Bristol.

Milborrow, D.J., 1980, The Performance of Arrays of Wind Turbines. Jnl Ind. Aerodynamics, 5 p.403-430.

Vermeulen, P.E.J., 1978, A Wind Tunnel Study of the Wake of a Horizontal Axis Wind Turbine, TNO report 78-09674.

Vermeulen, P.E.J., 1979, Mixing of Simulated Wind Turbine Wakes in Turbulent Shear Flow, TNO report 79-09974.

Vermeulen, P.E.J., 1980, An Experimental Analysis of Wind Turbine Wakes, Proc. 3rd Int. Symp. on Wind Energy Systems, Copenhagen, B.H.R.A., Cranfield.

Nomenclature

C_T Thrust coefficient

$$= \frac{\text{Thrust on wind turbine model}}{\frac{1}{2}\,\rho\,\pi\,R^2\,u_o^2}$$

D Diameter of wind turbine simulator

R Radius of wind turbine simulator

u_o Incident mean velocity at hub height

u Local mean velocity

u' Turbulent velocity

Z Vertical height

ρ Air density

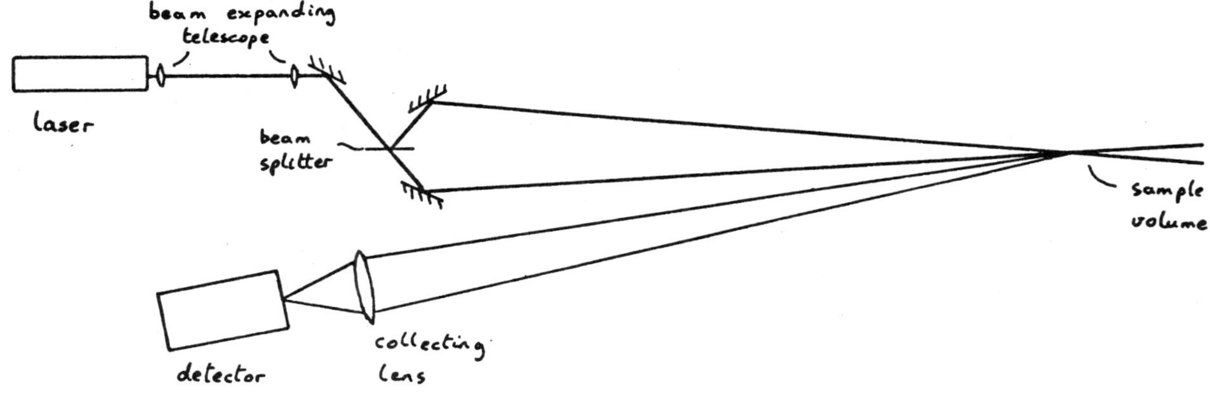

FIGURE 1: OPTICAL CONFIGURATION OF THE LASER DOPPLER ANEMOMETER

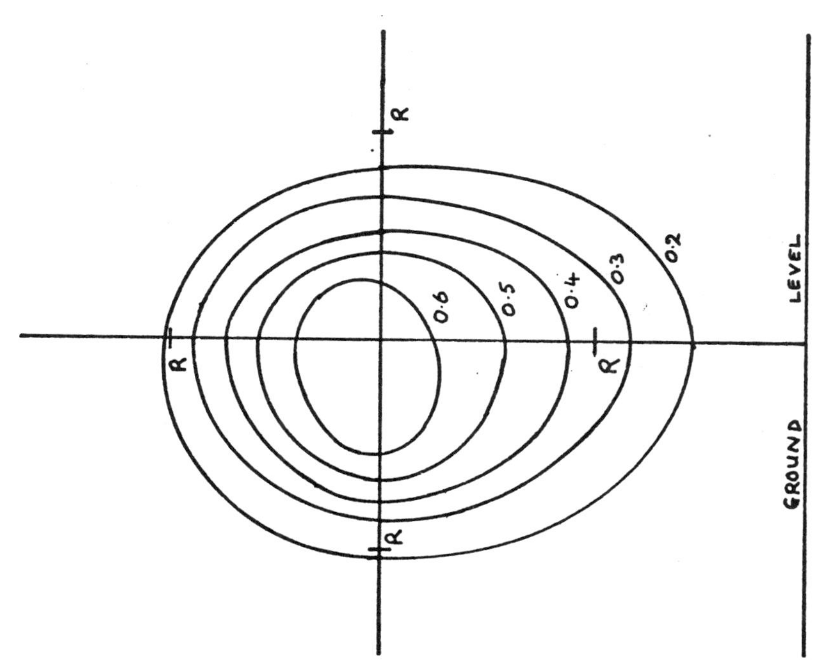

FIGURE 2: VELOCITY DEFICIT CONTOURS 2 DIAMETERS BEHIND THE SIMULATOR, THRUST COEFFICIENT 0.9

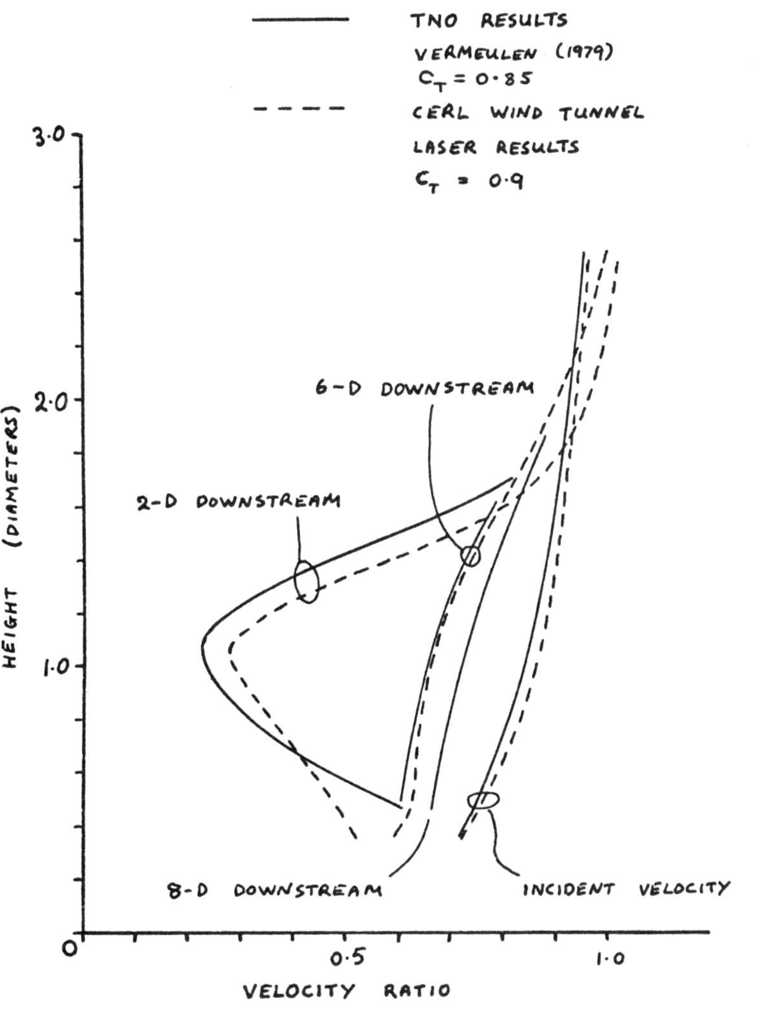

FIGURE 4: COMPARISON OF CERL VERTICAL WAKE PROFILES WITH TNO RESULTS

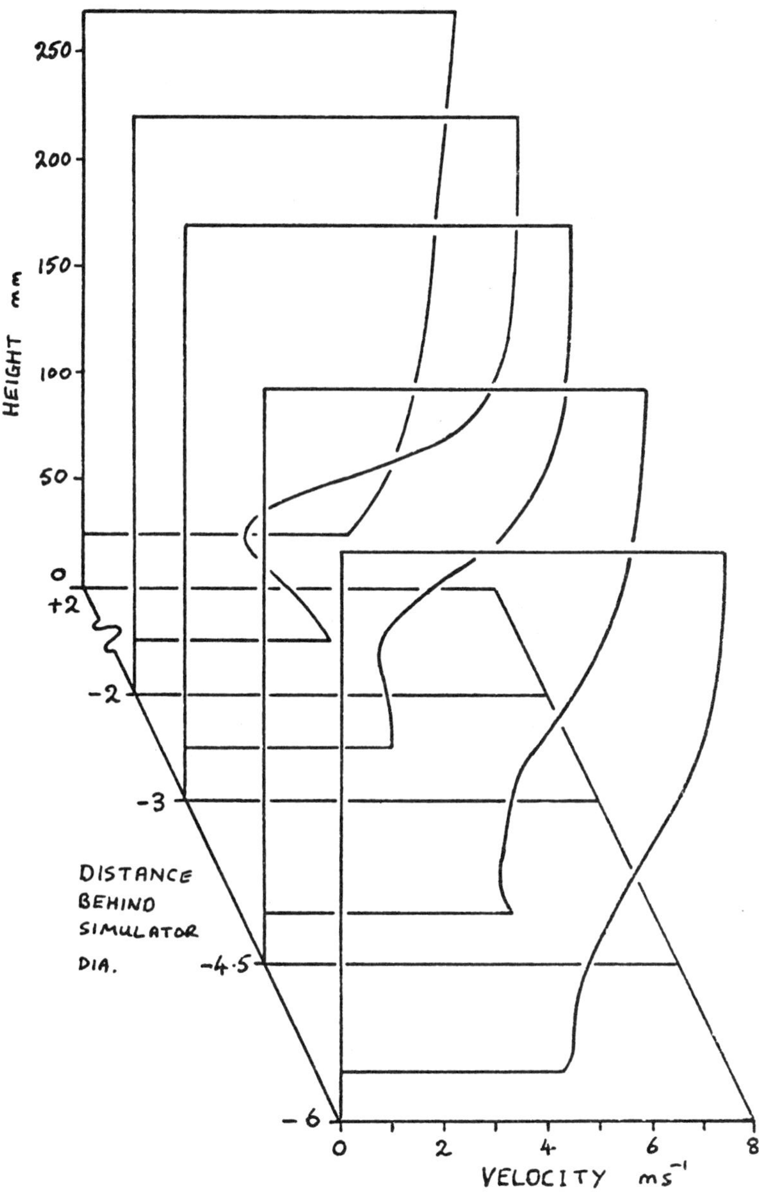

FIGURE 3: VERTICAL VELOCITY PROFILES BEHIND THE SIMULATOR, THRUST COEFFICIENT 0.9

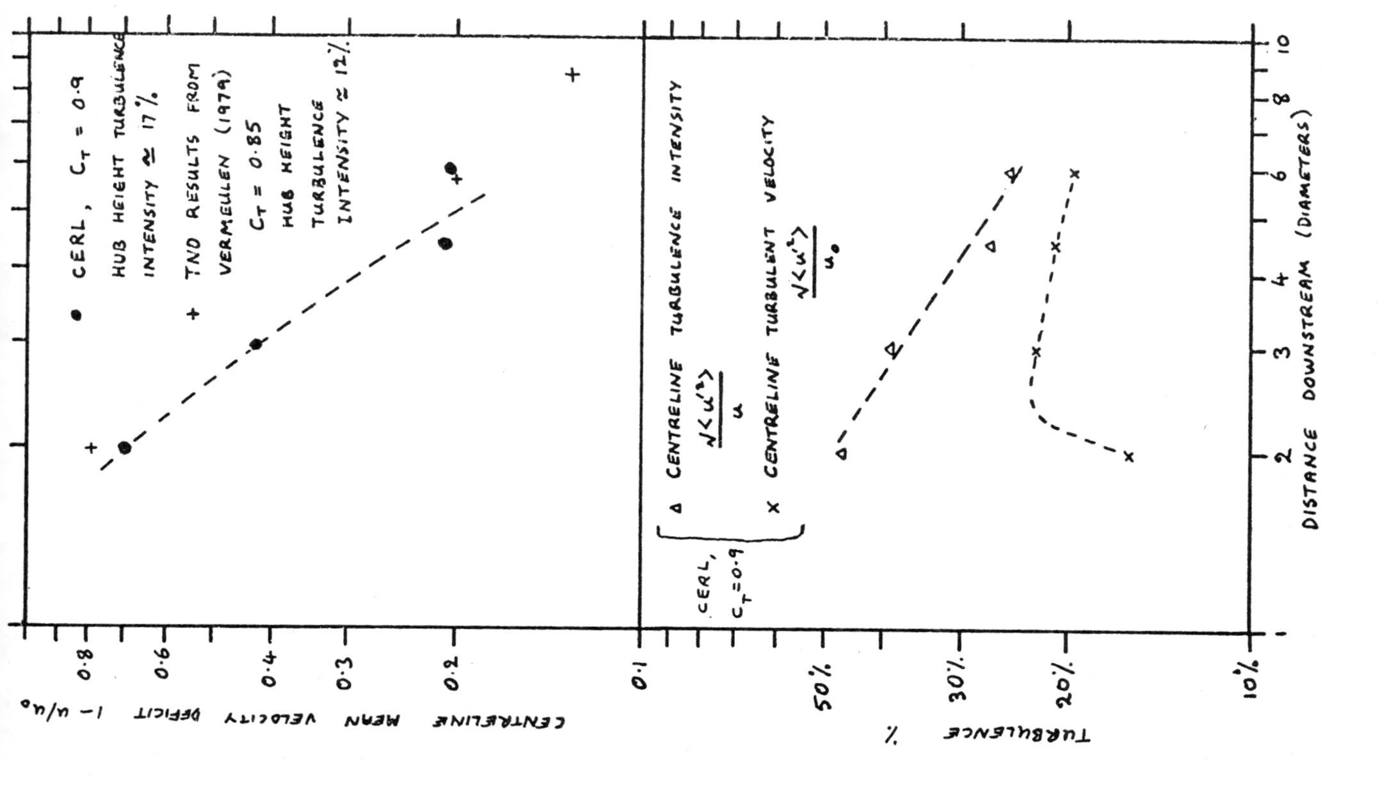

FIGURE 5: VARIATION OF MEAN VELOCITY AND TURBULENCE PARAMETERS ON THE CENTRELINE OF A SINGLE WAKE

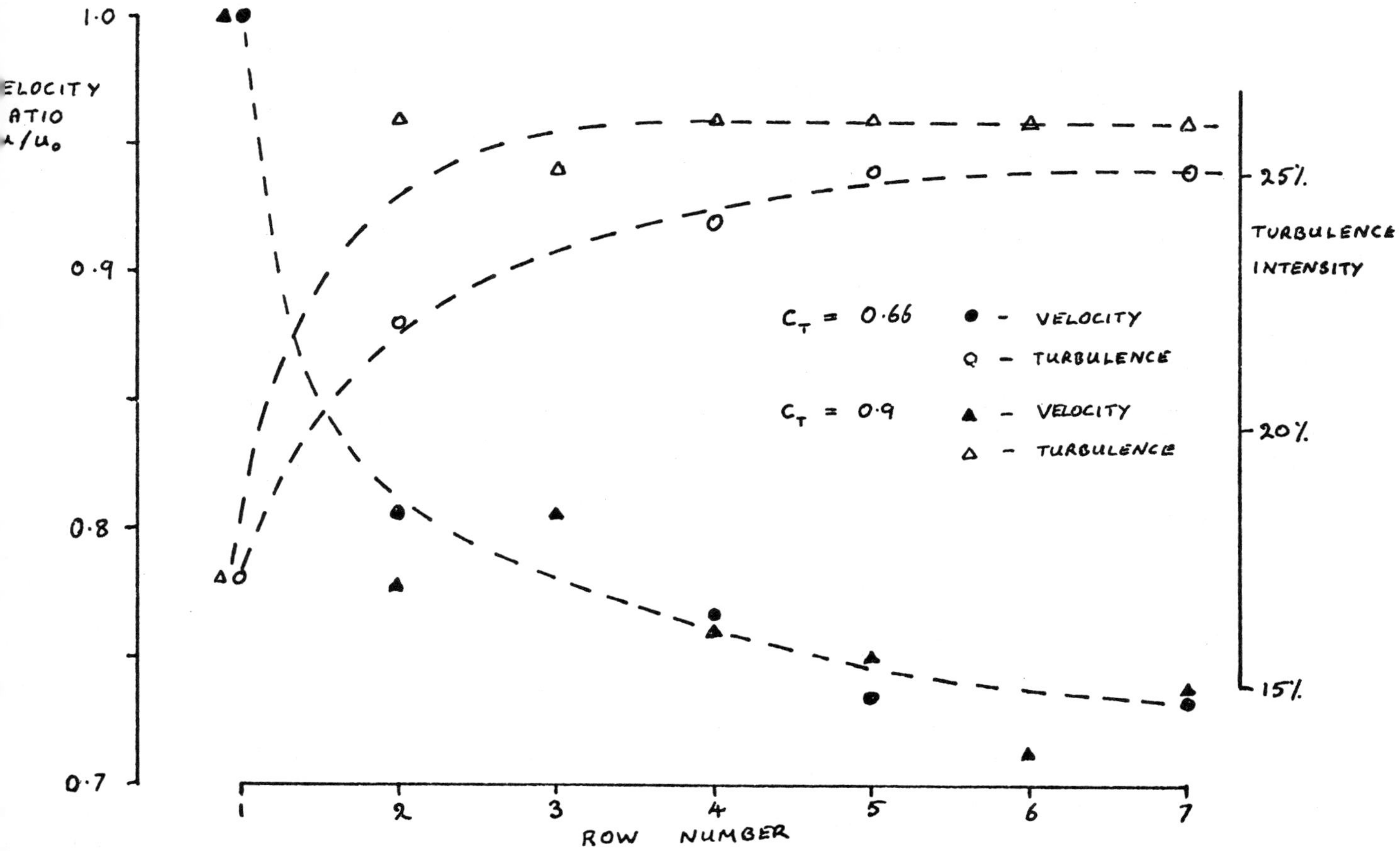

FIGURE 6: VELOCITY AND TURBULENCE INCIDENT ON EACH ROW OF THE CLUSTER AT HUB HEIGHT, (INCIDENT WIND DIRECTION 0°)

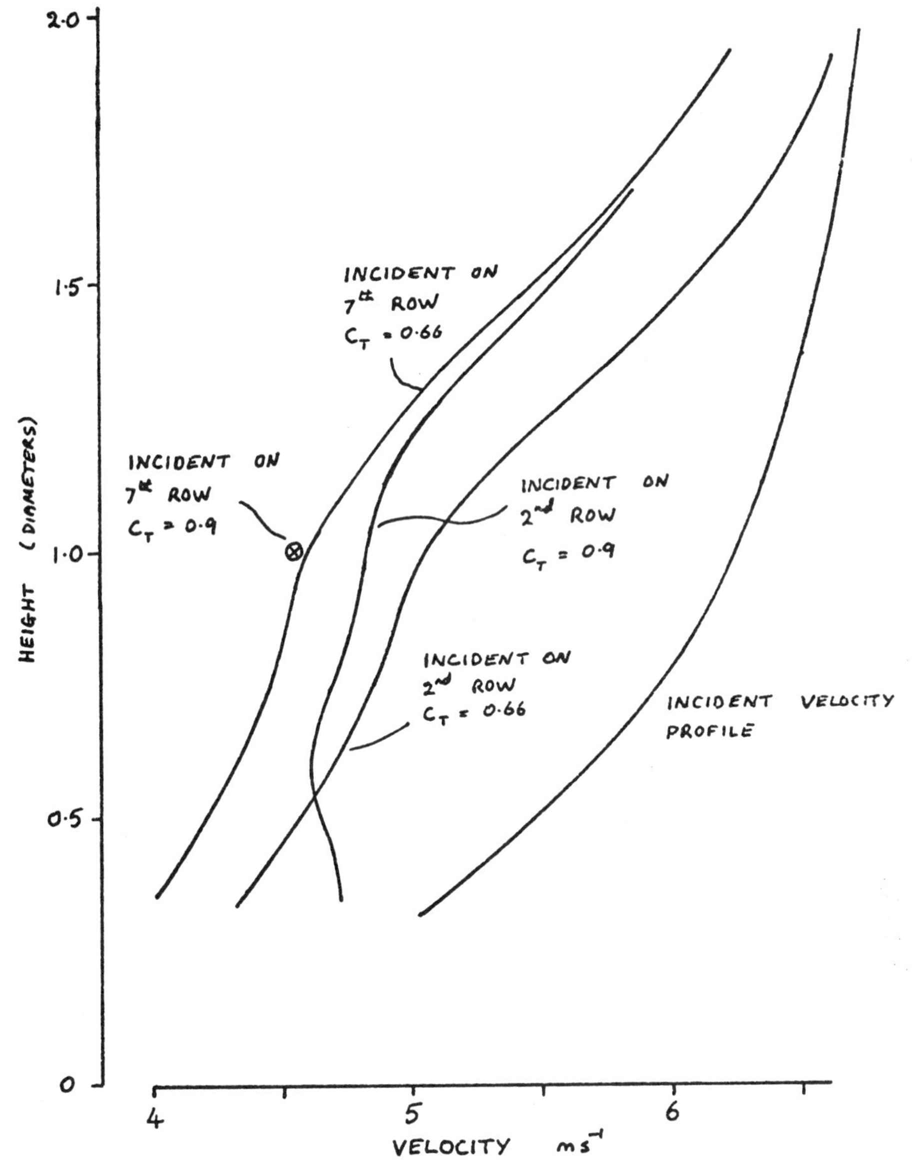

FIGURE 7: CHANGE IN BOUNDARY LAYER VELOCITY PROFILE THROUGH THE CLUSTER

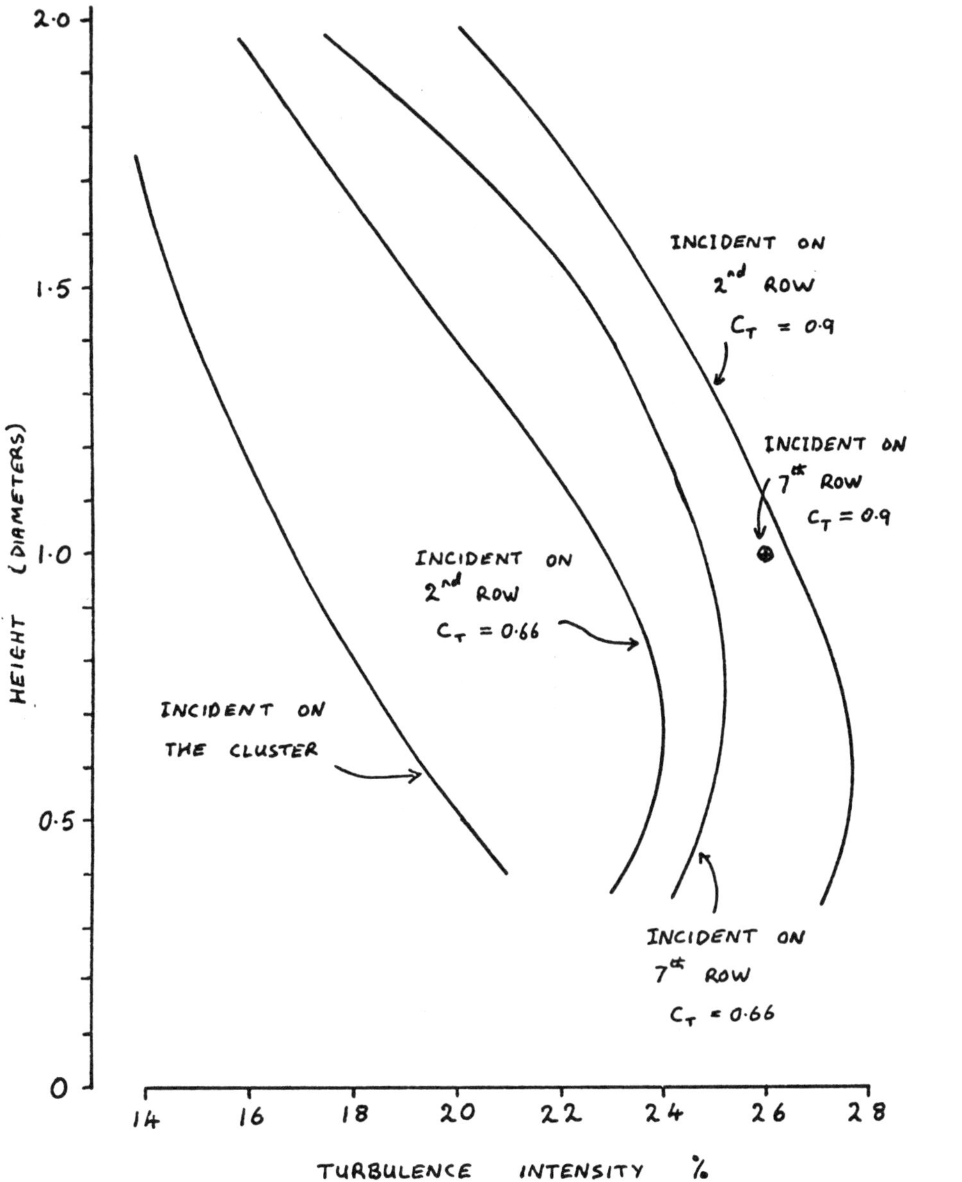

FIGURE 8: CHANGE IN TURBULENCE PROFILE THROUGH THE CLUSTER

THE EFFECT OF TERRAIN AND CONSTRUCTION METHOD ON THE FLOW OVER COMPLEX TERRAIN MODELS IN A SIMULATED ATMOSPHERIC BOUNDARY LAYER

D. Lindley[+], D. Neal[o], J.R. Pearse* and D.C. Stevenson[o]

+ Taylor Woodrow Construction Limited, Southall, U.K.
o University of Canterbury, Christchurch, New Zealand
* Shell, BP, Todd Oil Services, New Plymouth, New Zealand

Abstract

Details are given of measurements over a series of model two dimensional hills placed in a simulated 1:300 rural atmospheric boundary layer to determine the effect of slope, shape and surface roughness on mean wind velocity, RMS velocity fluctuations and energy spectra for the streamwise velocity component. Results of these tests are compared where possible with existing wind tunnel data.

The paper goes on to describe an extension of the work on two dimensional hills in which the flow over a 1:4000 undistorted scale model of Gebbies Pass in the South Island of New Zealand has been investigated in an atmospheric boundary layer wind tunnel. Three approaches to modelling complex terrain were employed. Velocity and turbulence profiles, Reynolds stresses and spectra have been measured over terraced, contoured and "roughness added" models and compared with field measurements taken in the modelled region. These results confirm the promise of similar experiments that physical modelling can make a useful and cost effective contribution in prospecting for and evaluating wind energy sites.

Introduction

Little quantitative information exists on the effect of hills on the local wind characteristics. A number of theoretical approaches to the problem of wind flow over hills have recently been made (Taylor and Gent (1974), Deaves (1975), Derickson and Meroney (1977)) but there is a notable absence of rigorous laboratory and field measurements with which to compare the theoretical models.

The tests described in the first part of this paper were concerned with investigating the influence of hill shape, hill slope and model hill surface roughness on the wind flow over two dimensional model hills. The work was performed in a model rural atmospheric boundary layer characterised by a one-sixth power law mean velocity profile and scaled about 1:300 between the atmosphere and the wind tunnel.

The second part of the paper concerns itself with an extrapolation of the technique to look at the flow over whole regions of complex terrain in which there may be one or more three dimensional hills. In so doing, it builds upon previous work by Meroney, Bowen, Lindley and Pearse (1978). There are still no satisfactory ways of prospecting for and evaluating potential wind energy sites without extensive and long term meteorological measurements. Physical simulation of wind regimes found over complex terrain offer significant advantages in terms of time, expense and control of independent variables.

Two Dimensional Hills

Seven two dimensional hills were investigated, four of the model hills were triangular in cross section with H/L equal to 0.10, 0.30, 0.50 and 1.00 and three model hills were bell shaped with H/L equal to 0.30, 0.50 and 1.00. H/L is defined as the ratio of the maximum height of the hill to the distance from the point of maximum height to where the hill height is half its maximum. This definition of L gives a hill slope of 0.5 H/L for triangular ridges. The three bell shaped hills had cross sections defined by

$$f\left(\frac{x}{L}\right) = \frac{H}{\left[1 + \left(\frac{x}{L}\right)^2\right]}$$

which is a geometry used in the theoretical approaches of Corby (1954) and Jackson and Hunt (1975) and is a good approximation to the bell shaped hills considered by Taylor and Gent (1974), Deaves (1975) and Derickson and Meroney (1977).

All seven model hills had a smooth surface in contrast to the 3 mm high uniform roughness covering the floor of the 12 m long working section of the boundary layer wind tunnel. To investigate the effect of the model hill surface roughness, tests were performed using two additional triangular hills with H/L equal to 0.10 and 1.00 covered with the roughness used on the floor of the wind tunnel.

The mean velocity profile was characteristic of neutrally stable atmospheric flow over rural terrain of roughness length Zo about 50 mm linearly scaled by a factor of approximately 1:300. A complete description of the experiment which includes details of approach mean velocity profiles, u component of turbulence intensity and energy spectra for the u velocity component have been given elsewhere by Pearse (1979) and Pearse, Lindley and Stevenson (1981).

Effect of the Hill Slope

The lines of equal amplification factor plotted in Figure 1 for each of the four smooth triangular hills in the 1:300 rural atmospheric boundary layer simulation show that upstream of the hill crests the maximum reduction in velocity occurred at the foot of the hill. As the flow moved along the upstream slope the amplification factor increased rapidly. The amplification factor is defined as the mean wind speed at a height z above ground level divided by the reference mean wind speed which is measured at the same height above ground level but in the undisturbed flow over flat ground upstream of the hill.

The maximum velocity over each hill was measured at the hill crest and the maximum crest velocity occurred for the hill with H/L of 0.50. The steepest hill did not have the largest amplification factors because the upstream vortex and downstream wake region effectively made the hill of a more gentle slope.

The amplification factors at the crest of the triangular hills are compared to those measured by Bouwmeester et. al. (1978) in Figure 2. Considering the slightly different ratios of H/δ and H/z_0 in each work, the agreement between the two sets of results is good. Comparison is also made in Figure 2 with the potential flow solution presented by Jackson (1979). Jackson's solution agrees well with the experimental results except for the steepest hill (H/L = 1.0) where separated flow occurs. For the hill with H/L = 0.1, better agreement with theory is obtained for the hill covered with roughness as the flow does not speed up due to the surface roughness change.

The maximum value of A_z of 1.82 at the crest of the hill with H/L of 0.50 at the lowest height investigated (z/H = 0.2) was the same as the peak value of A_z at the same height measured by Bowen and Lindley (1977) over their rough escarpment shapes. This is in contrast to the potential flow theory presented by Jackson (1979). Jackson showed that the increase in velocity at the crest for an escarpment should be half that for a hill of the same slope. The steep hill slope and the consequent flow separation conflict with the assumptions of the theory. In addition, the flow passing over the smooth model hill surface would have speeded up due to the change in surface roughness.

After the hill crest the mean velocity decreased for all four hills. A wake region characterised by low mean velocities and high turbulence intensities grew after the crest for the hill with H/L values of 0.50 and 1.00. The more gentle slopes for the hills with H/L of 0.10 and 0.30 precluded the occurrence of a downstream wake region.

The profiles of velocity fluctuations in Figure 3 illustrate the small decrease in u' at the hill crest as predicted by rapid distortion theory. For those hills which exhibited a wake region, high values of u' occurred in the post crest region. The limitations of hot wire anemometers in wake regions are well known and the measurements in these areas are not accurate and only indicate trends.

The spectral shape was invariant as the flow moved over the hills, but the spectral peak wave number decreased at the hill crest for all the hills as shown in Table I. The largest decrease occurred for the hills with H/L of 0.30 and 0.50 which also exhibited the largest increase in mean velocity. Comparison of the spectral peak wave number for the rough and smooth triangular hills with H/L = 0.10 in Table 1 suggests that the surface roughness change associated with the smooth model hills may have caused the decrease in the spectral peak wave number at the hill crests.

Effect of the Hill Surface Roughness

Lines of equal amplification factor for the flow over the rough surfaced triangular hills with H/L equal to 0.10 and 1.00 in the 1:300 rural atmospheric boundary layer simulation are shown in Figure 4. Comparison of Figure 4 with Figure 1 illustrates the much larger effect of the hill surface roughness on the hill with H/L of 0.10 which had the longer hill surface. The increased surface roughness caused the velocity to decrease over the hill.

In an effort to separate the influences of topography and surface roughness changes, mean velocities were measured over a 2 m length of flat acrylic sheet. The acrylic sheet was placed in the wind tunnel working section so it had the same upstream reference profile as the model hills.

The speed-up due to the flow over the 2 m length of flat sheet was calculated. A 2 m length of flat acrylic sheet was chosen as this corresponded to the length of hill surface for the triangular hill with an H/L value of 0.10. The speed-up calculated for the flow over the flat sheet is compared with the amplification factors obtained for the rough and smooth triangular hills with H/L equal to 0.10 in Table II.

In comparing the amplification factors over the flat sheet with those over the hill surface, it was assumed that there would be no pressure effect from the change in surface roughness alone and that the hill slope by itself had no effect on the stress distribution except for displacing it along with the flow. This may not be quite correct since a change in the roughness of the surface may have at least some minor effect on the pressure distribution.

The close correspondence between the two columns on the right hand side of Table II suggests that the effect of the smooth hill surface is exactly the same as the effect of the same length of flat smooth surface, thus confirming the principle of superposition as suggested by Jensen and Peterson (1978).

The turbulence profiles at the hill crests, presented in Figure 3, show that the surface roughness over the hill with H/L equal to 0.10 caused the streamwise velocity fluctuations to increase compared to the measurements made over the hill with a smooth surface. The largest changes occurred close to the ground. For the steeper hill with H/L equal to 1.00, the changes in the turbulent fluctuations caused by the different hill surface roughness were within the experimental measurement uncertainty.

The decrease in the mean velocity and the corresponding increase in the velocity fluctuations over the model hill caused by the rougher hill surface were due to the surface protrusions over the hill surface causing a larger drag and a more turbulent eddying motion. This more turbulent flow was at a higher wave number as indicated by the spectral peak wave numbers for the rough and smooth hills in Table I.

Effect of the Hill Shape

Comparison of the amplification factors over the bell-shaped hills in Figure 5 and the triangular hills in Figure 1 for the same value of H/L shows that the bell-shaped hills had much larger velocities before and after the hill crest due to their less abrupt start and finish, compared to the triangular hills. The amplification factors at the crest of the triangular and bell-shaped hills compared in Figure 6, show that the bell-shaped hills had slightly larger amplification factors at the hill crest compared to the triangular hills.

Bouwmeester et. al. (1978) compared mean velocity measurements over triangular and half-sinusoidal hills with H/L values of 0.50 and 0.67, and found that the differences in the mean velocities over the model hills were small.

The velocity fluctuations u' at the hill crests were about 10% lower for the bell-shaped hill with H/L equal to 0.50 than for the corresponding triangular hill, but were within the experimental measurement uncertainty for the hills of aspect ratio 0.10 and 1.00.

The spectral peak wave numbers presented in Table I show that there was a larger decrease in the peak wave number at the crest as the flow moved over the bell-shaped hills than over the triangular hills.

Considering the changes in the mean velocity, velocity fluctuations and spectra, the model hill shape does not affect the flow as much as the hill slope and the hill surface roughness.

Conclusions

Flow measurements over smooth triangular hills of four different slopes showed that the greatest increase in the mean velocity over the model hills occurred at the crest for a hill of intermediate slope rather than for the steepest hill. There was little change in the values of the velocity fluctuations except in the wake regions which formed downstream of the steeper hills.

The shape of the model hills was shown to be important only for those hills with a slope near that slope at which a wake region formed. For such cases the round crested hills delayed the onset of formation of a wake region.

The effect of the surface roughness change associated with a simulated atmospheric boundary layer developed over a rough surface passing over a smooth hill surface was examined by comparing the measurements made over two hills of the same shape with rough and smooth surfaces. The hill surface roughness was found to have a significant effect on the flow quantities over the more gentle hills causing changes of up to 20% in the mean velocity close to the ground.

Complex Terrain Model

Complete simulation of wind characteristics over irregular terrain implies that the two flow systems are thermally, dynamically and kinematically similar. Meroney et. al. (1978) have discussed the problem of obtaining similarity and suggest that to obtain surface boundary similarity, the following are required:

(a) surface roughness distribution with aerodynamically rough behaviour

(b) topographic relief

(c) surface temperature distribution

and for approach flow characteristics similarity, the following are required:

(i) distribution of mean and turbulent velocities

(ii) distribution of mean and fluctuating temperature

(iii) zero longitudinal pressure gradient

(iv) if the flow is thermally layered, equality of the ratio of inversion depths is required.

One of the requirements suggested by Meroney et. al. (1978) for fully rough flow was:

$$\frac{u_* \lambda}{\nu} \quad 100$$

where u_* = friction velocity
ν = kinematic viscosity
λ = roughness element height (the polystyrene sheet used was assessed to be 0.05 to 0.2 mm)

Taking x = 0.1 yields $\frac{u_* \lambda}{\nu}$ $0(10^3)$

Values for the Gebbies Pass model taken at the gradient height and 3 mm above the surface area are $(0)10^5$ and $(0)10^4$ respectively.

Laboratory Measurements

The area considered in this part of the study and shown in Figure 7 is located on Banks Peninsula in New Zealand's South Island. The Gebbies Pass saddle lies at right angles to the prevailing winds from the north-east and south-west.

Figure 7 shows Gebbies Saddle to be bounded on one side by the Port Hills which rise to above 600 m. These are generally sparsely covered with vegetation other than grass. On the other lies Mt. Herbert Range generally about 500 m above sea level, whilst the Saddle itself is approximately 300 m above sea level. The approaching wind flow from the sea passes over flat grass-covered terrain on which are several shelter-belt stands ranging from thick hedges 5-10 m high to conifer trees 10-20 m high. The flow passes over approximately 3 km of this terrain before entering the modelled region.

Model Construction

The model was constructed from 9 mm thick expanded polystyrene-bead board cut to the shape of the terrain contours of a photographically enlarged lands and Survey Department map, thus producing a terraced finish to the model. A non-shrink polyfilla was used to fill in the terraces to produce the contoured model. The final form of construction included the modelling of surface roughness elements on the model. The shelter belts were simulated as in the experiments done by Meroney et. al. with an open spun wool approximately 4 mm thick whilst the scrub areas, typically 3-5 mm high, were represented by an open weave hessian providing a close physical model in terms of height and density.

The model was surveyed in the wind tunnel in all three construction forms, i.e. terraced, contoured and with roughness elements added.

Simulated Atmospheric Boundary Layer

The measurements over the model were made in the same 1.2 m x 1.2 m cross section boundary layer wind tunnel used in the first series of experiments on two dimensional hills.

Measurements of the velocity and turbulence intensity profiles, roughness length and the longitudinal componenet of the energy spectrum up to a height of 20 m were made on the approach terrain with a field data acquisition system incorporating fast response propeller anemometers, and data recording facilities. Details of this system have been given by Flay (1978).

The results obtained with this equipment combined with those available in the literature were used in establishing the correct approach flow characteristics in the simulated wind tunnel boundary layer.

A variety of flow visualisation techniques including polystyrene beads and flags were used in combination with detailed surveys using a 5 hole cobra probe and hot film anemometers to yield details of the flow field.

Field Measurement Programme

Wind velocity measurements over the Gebbies Pass region were obtained to provide a basis for comparison with those made over the physical model. In the absence of long term velocity and direction data, a mobile survey technique, similar to that used by Meroney et. al. was employed. This involved teams covering the area in periods of 5 to 7 hours whilst making measurements with a lightweight 10 m aluminium guyed pole and a Rimco cup anemometer driving a counter. Five minute means of velocity so obtained were normalised using a wind speed record obtained from a tower situated upstream of the complex terrain. Altogether, measurements were obtained on six such field days.

Results and Conclusions

Complete details of this study have been given by Neal (1979) and reported by Neal, Stevenson and Lindley (1981). Only a brief summary of the results will be given here. The flow visualisation studies showed that the portion of the model shown in Figure 8 was not subjected to major flow direction changes. The points analysed on the model formed the grid pattern shown in this Figure, thus allowing the study of the flow through longitudinal and lateral cross sections over the model.

Iso contour diagrams were prepared from the laboratory velocity and turbulence intensity measurements. These took the form of isotach and isoturb contour plots, such as those given for the terraced and roughness added form of construction for longitudinal cross sections YY given in Figures 9 and 10.

The results obtained with the model in the terraced form of construction were markedly different to those for the contoured and roughness added models. In particular, the terraces produced lower velocities and higher turbulence intensity values on the windward face of hills and ridges, as might have been expected.

The general terrain configuration showed, by means of speed up ratio plots, that the wind velocity increased significantly (50-80%) as it passed over the Saddle. The major conclusions arising from the study are:

1. Approach flow characteristics could be simulated correctly, these being confirmed by field measurements on the prototype.

2. Accurate reproduction of surface roughness and terrain shape were required to produce equivalent wind speeds close to the ground i.e. in the bottom 100m of the boundary layer.

3. Terraced models were shown to be unsatisfactory for WECS site selection surveys. However,

4. Physical modelling reproduced the relative wind speeds in a complex terrain situation to produce sample correlation co-efficients of 0.71 to 0.97 thus being similar to the range of variability found between the data from the various field days.

References

1. Bouwmeester, R.J.B., Meroney, R.N. and Sandborn, V.A. (1978). Sites for wind power installations: Wind Characteristics over Ridges. Report No. CER 77-78 RJ BB - RNM - VAS51, Fluid Mechanics and Wind Engineering Program, Colorado State University, Fort Collins, Col.

2. Bowen, A.J. and Lindley, D (1977). A wind tunnel investigation of the wind speed and turbulence characteristics close to the ground over various escarpment shapes, Boundary Layer Meteorol, 12, pp. 259-71.

3. Corby, G.A. (1954). The airflow over mountains. A review of the state of current knowledge. Quart, J.Roy. Meteorol. Soc., 80, 491-521.

4. Deaves, D.M. (1975) Wind over Hills: A numerical approach. Journal of Indust. Aero. 1, 371-391.

5. Derickson, R.G. and Meroney, R.N. (1977). A simplified Physics airflow model for evaluating wind power sites in complex terrain. Proc. Summer Computer Simulation Conference, Chicago, USA. July 1977, 14 pp.

6. Flay, R.G.J. (1978) Structure of a rural atmospheric boundary layer near the ground. Ph.D Thesis, Dept. of Mechanical Engineering, University of Canterbury, Christchurch, New Zealand.

7. Jackson, P.S. and Hunt, J.C.R. (1975) Turbulent wind flow over a low hill - Quart, J. Roy. Meteorol. Soc., 101, pp. 929-955.

8. Jackson, P.S. (1979) The influence of local terrain features on the site selection for wind energy generating systems. Report BLWT - 1 - 79, Boundary layer wind tunnel Lab., University of Western Ontario, London, Ontario, Canada.

9. Jensen, N.O. and Petersen, E.W. (1978) On the escarpment wind profile. Quart, J. Roy, Meteorol. Soc., 104, pp. 719-28.

10. Meroney, R.N., Bowen, A.J., Lindley, D. and Pearse, J.R. (1978) Wind characteristics over complex terrain: Laboratory Simulation and field measurements at Rakaia Gorge, New Zealand. U.S. Dept. of Energy Report RLO - 2438 - 77/2 May. NTIS U.S. Dept. of Commerce, Springfield, Virginia.

11. Neal, D. (1979) Wind flow and structure over Gebbies Pass, New Zealand. A comparison between a wind tunnel simulation and field measurements. Ph. D Thesis, Department of Mechanical Engineering, University of Canterbury, Christchurch, New Zealand.

12. Neal, D, Stevenson, D.C., and Lindley, D (1981) A wind tunnel boundary Layer simulation of wind flow over complex terrain. Effect of terrain and model construction. Boundary Layer Meteorology (to be published).

13. Pearse, J.R. (1979) The influence of two-dimensional hills on simulated boundary layers - Vols. 1 and 2. Ph. D Thesis, Department of Mechanical Engineering, University of Canterbury, Christchurch, New Zealand.

14. Pearse, J.R., Lindley, D., and Stevenson, D.C. (1981) Wind flow over ridges in simulated atmospheric boundary layers. Boundary layer meteorology (to be published).

15. Taylor, P.A. and Gent, P.R. (1974) A model of Atmospheric Boundary Layer flow above an isolated two-dimensional hill; an example of flow above "Gentle Topography". Boundary Layer Meteorology, 7, 349-362.

	$\frac{H}{L}$	$\frac{Z}{H} = 0.2$						$\frac{Z}{H} = 1.0$						$\frac{Z}{H} = 2.0$					
		10H up-stream of hill		Hill Crest		10H down-stream of hill		10H up-stream of hill		Hill Crest		10H down-stream of hill		10H up-stream of hill		Hill Crest		10H down-stream of hill	
		$\left(\frac{n}{\bar{U}}\right)_p$	n_p	$\left(\frac{n}{\bar{U}}\right)_p$	n_p	$\left(\frac{n}{\bar{U}}\right)_p$	n_p	$\left(\frac{n}{\bar{U}}\right)_p$	n_p	$\left(\frac{n}{\bar{U}}\right)_p$	n_p	$\left(\frac{n}{\bar{U}}\right)_p$	n_p	$\left(\frac{n}{\bar{U}}\right)_p$	n_p	$\left(\frac{n}{\bar{U}}\right)_p$	n_p	$\left(\frac{n}{\bar{U}}\right)_p$	n_p
		m^{-1}	Hz	m^{-1}	Hz	m^{-1}	Hz	m^{-1}	Hz	m^{-1}	Hz	m^{-1}	Hz	m^{-1}	Hz	m^{-1}	Hz	m^{-1}	Hz
Smooth Triangular Hills	0.1	0.82	6.36	0.28	3.56	0.96	7.90	0.37	4.09	0.30	4.03	0.39	4.63	0.35	4.29	0.34	4.76	0.34	4.25
	0.3	0.82	6.36	0.29	4.07	1.00	8.15	0.37	4.09	0.27	3.69	0.34	3.92	0.35	4.29	0.29	4.09	0.29	3.66
	0.5	0.82	6.36	0.25	3.48	1.60	12.32	0.37	4.09	0.27	3.76	0.35	3.68	0.35	4.29	0.29	4.05	0.29	3.47
	1.0	0.82	6.36	0.34	4.30	2.90	15.89	0.37	4.09	0.33	4.47	1.10	8.72	0.35	4.29	0.33	4.57	0.70	7.67
Rough Triangular Hills	0.1	0.82	6.36	0.80	8.50	1.30	9.98	0.37	4.09	0.40	5.33	0.44	4.88	0.35	4.29	0.35	4.85	0.37	4.83
	1.0	0.82	6.36	0.42	5.35	1.70	10.56	0.37	4.09	0.34	4.69	1.40	12.54	0.35	4.29	0.27	3.75	0.55	6.47
Bell Shaped Hills	0.3	0.82	6.36	0.27	3.74	0.82	6.57	0.37	4.09	0.27	3.73	0.36	4.20	0.35	4.29	0.28	3.95	0.28	3.53
	0.5	0.82	6.36	0.17	2.61	1.00	7.92	0.37	4.09	0.24	3.47	0.42	4.89	0.35	4.29	0.28	4.02	0.32	4.00
	1.0	0.82	6.36	0.29	3.78	1.18	8.08	0.37	4.09	0.27	3.66	0.67	6.32	0.35	4.29	0.28	3.91	0.46	5.00

TABLE I: Spectral Peak Frequencies for the u velocity component

Height $\frac{z}{H}$	Amplification factor over the hill $\frac{H}{L} = 0.1$ at $\frac{x}{L} = 2$		Difference in amplification factors	Speed-up due to 2m of flat acrylic sheet
	No surface roughness	Surface roughness		
0.2	1.17	0.87	0.30	0.31
0.4	1.10	0.92	0.18	0.16
0.6	1.06	0.92	0.14	0.10
0.8	1.04	0.97	0.07	0.07
1.0	1.03	0.97	0.06	0.06
1.5	1.01	0.99	0.02	0.03
2.0	1.01	1.00	0.01	0.02
3.0	1.01	1.00	0.01	0.00
4.0	1.01	1.03	0.02	0.01
5.0	1.01	1.02	0.01	0.01
7.0	1.01	1.01	0.00	0.00
10.0	0.99	1.01	0.02	0.00
13.0	1.00	1.01	0.01	0.00

TABLE II: Comparison of the Amplification Factors over a Flat Surface and over a Rough and Smooth Hill

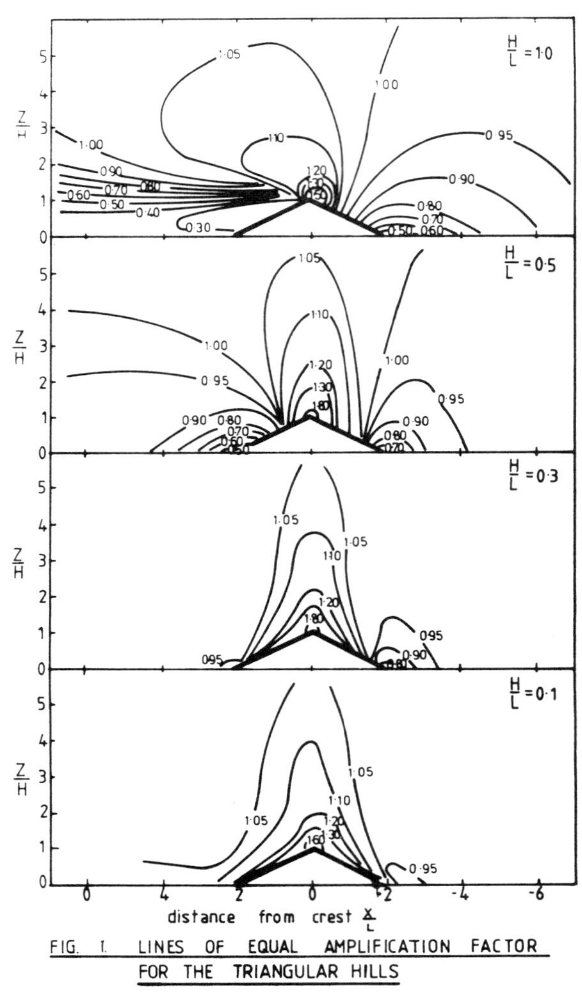

FIG. 1. LINES OF EQUAL AMPLIFICATION FACTOR FOR THE TRIANGULAR HILLS

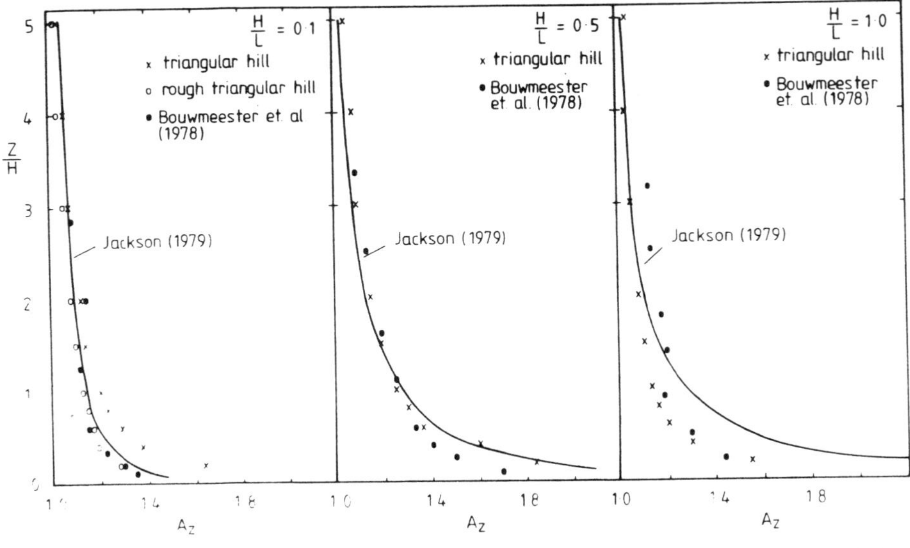

FIG 2 AMPLIFICATION FACTORS AT THE CREST OF THE SMOOTH TRIANGULAR HILLS IN THE 1:300 RURAL SIMULATION

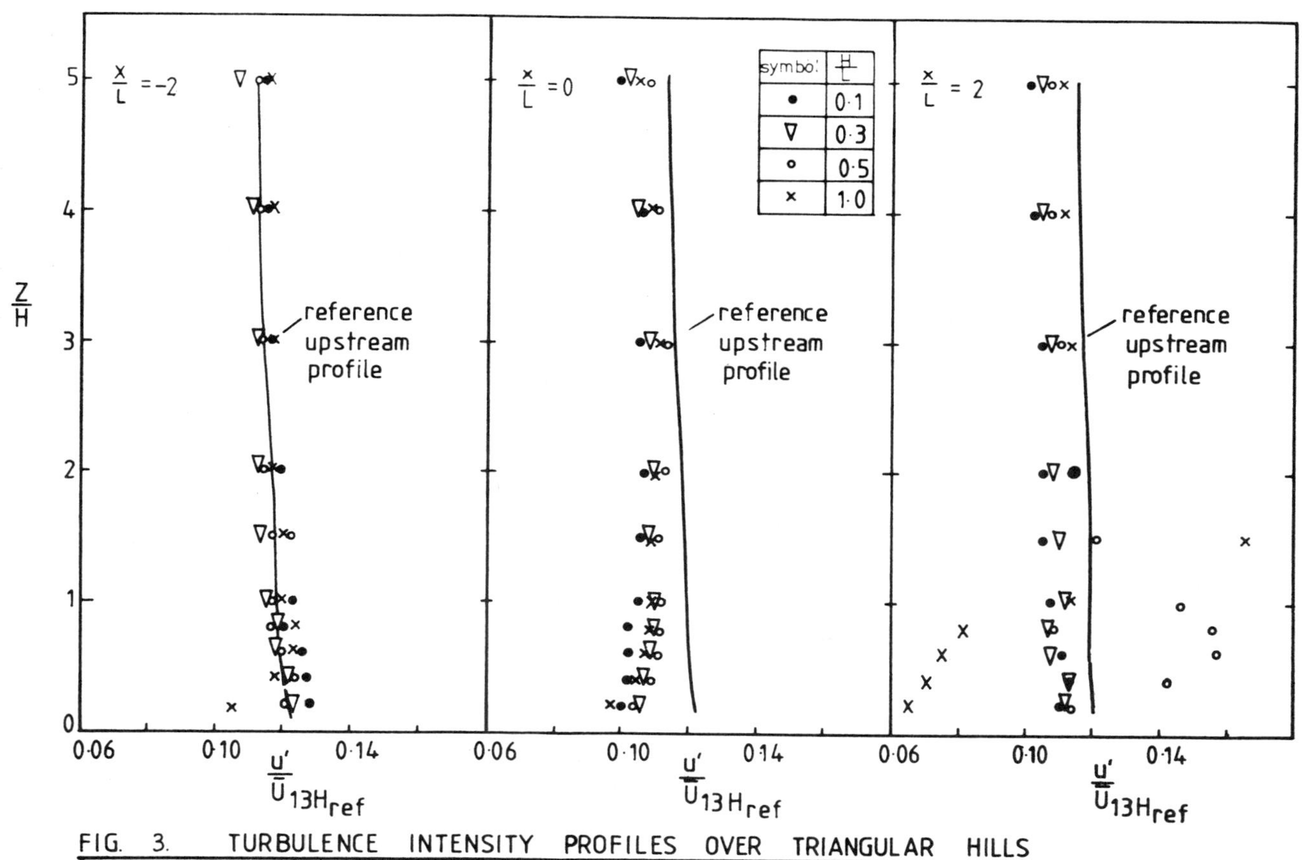

FIG. 3. TURBULENCE INTENSITY PROFILES OVER TRIANGULAR HILLS

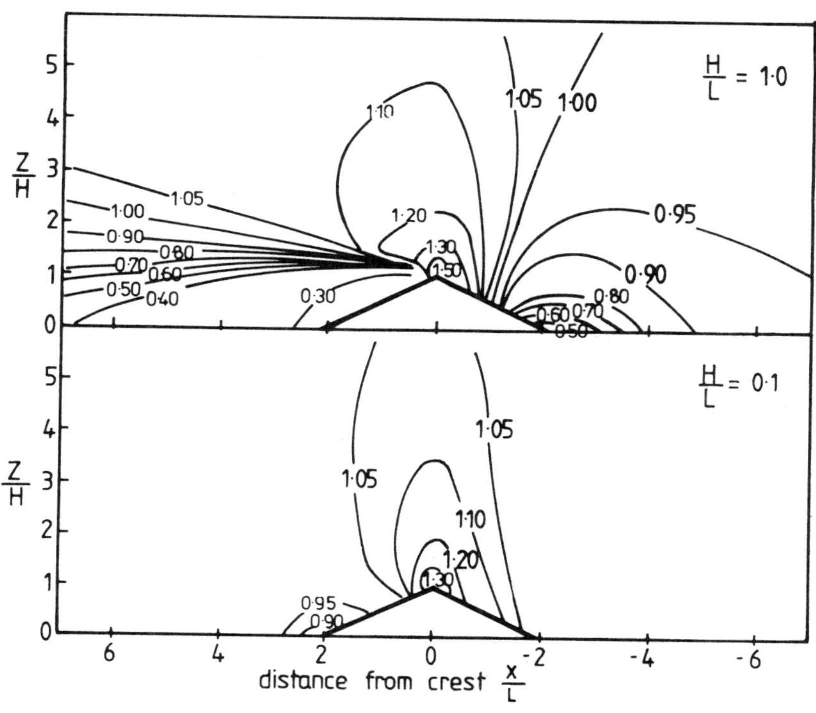

FIG. 4. LINES OF EQUAL AMPLIFICATION FACTOR OVER THE ROUGH TRIANGULAR HILLS

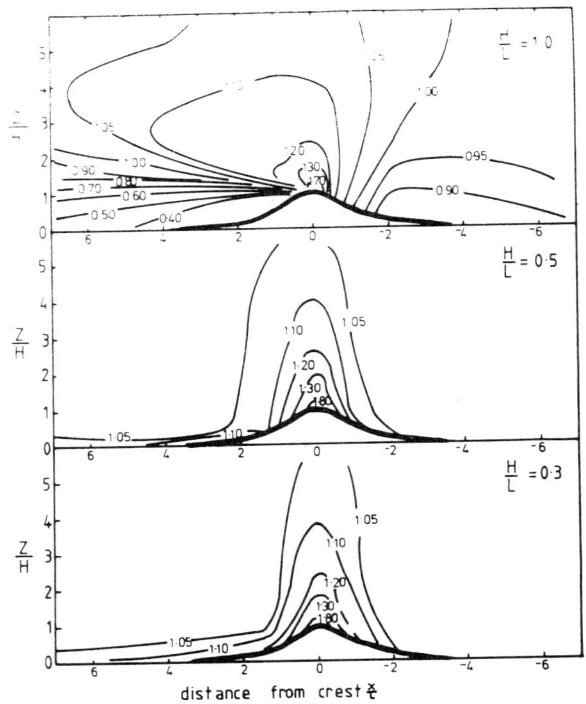

FIG. 5. LINES OF EQUAL AMPLIFICATION FACTOR FOR THE BELL SHAPED HILLS

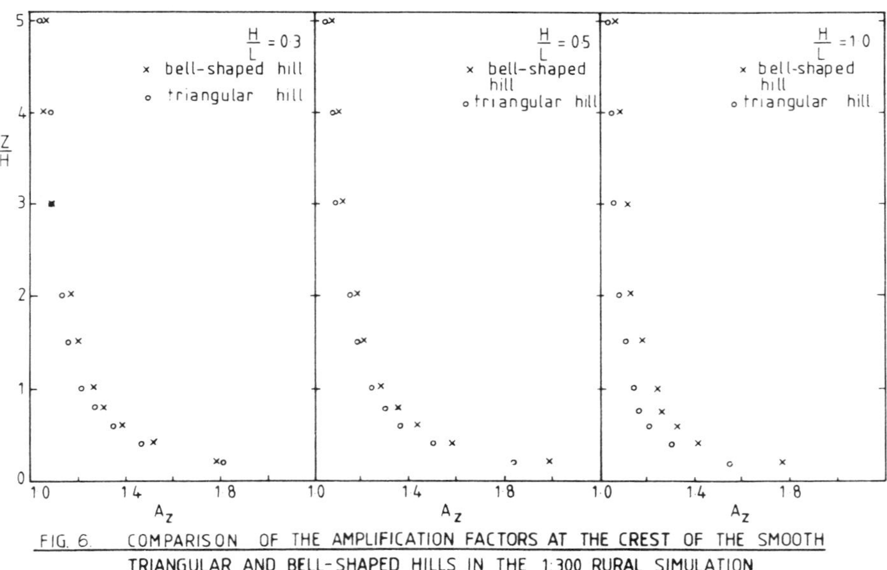

FIG. 6. COMPARISON OF THE AMPLIFICATION FACTORS AT THE CREST OF THE SMOOTH TRIANGULAR AND BELL-SHAPED HILLS IN THE 1:300 RURAL SIMULATION

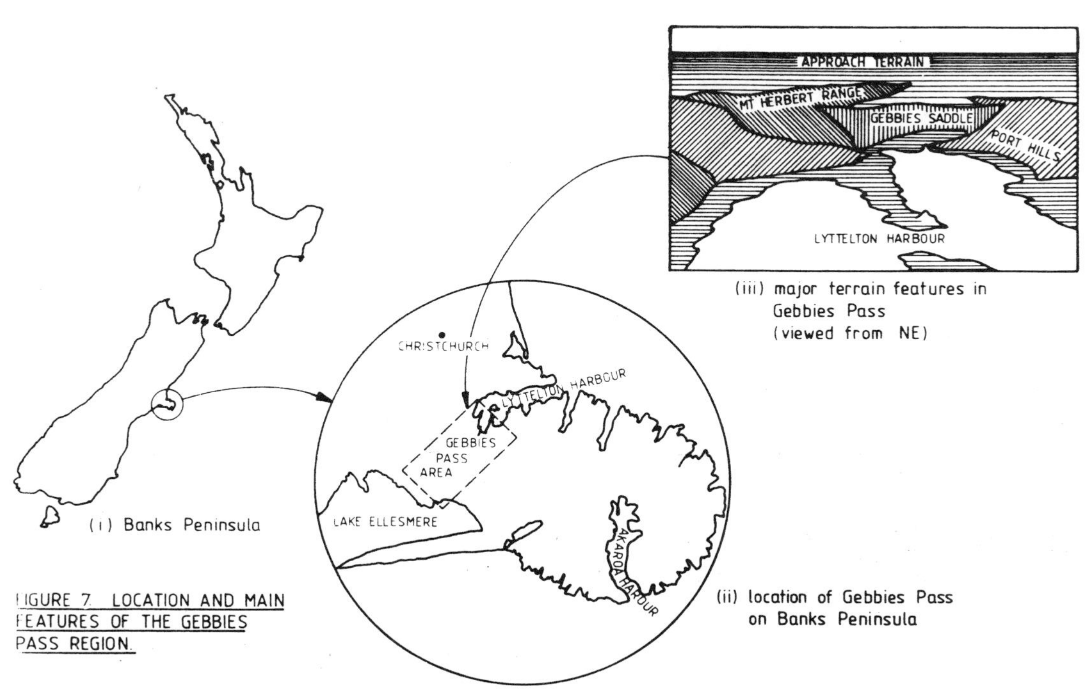

FIGURE 7. LOCATION AND MAIN FEATURES OF THE GEBBIES PASS REGION.

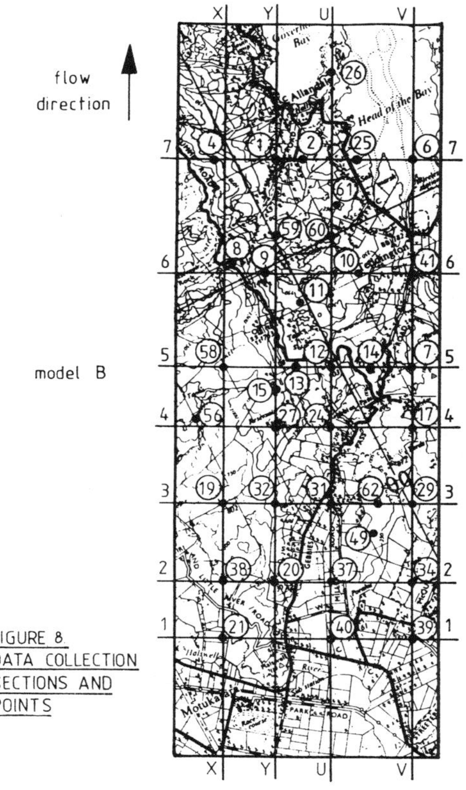

FIGURE 8. DATA COLLECTION SECTIONS AND POINTS

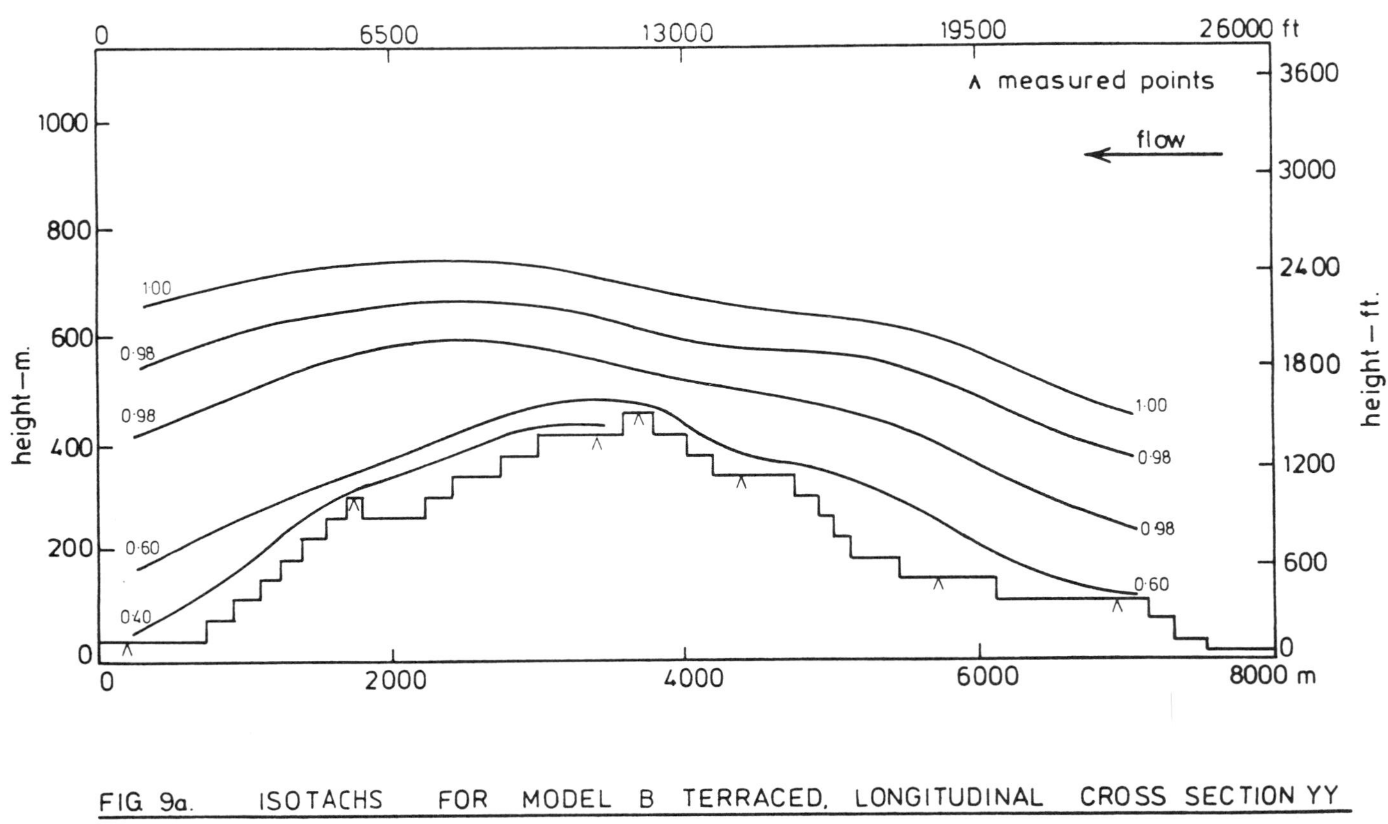

FIG 9a. ISOTACHS FOR MODEL B TERRACED, LONGITUDINAL CROSS SECTION YY

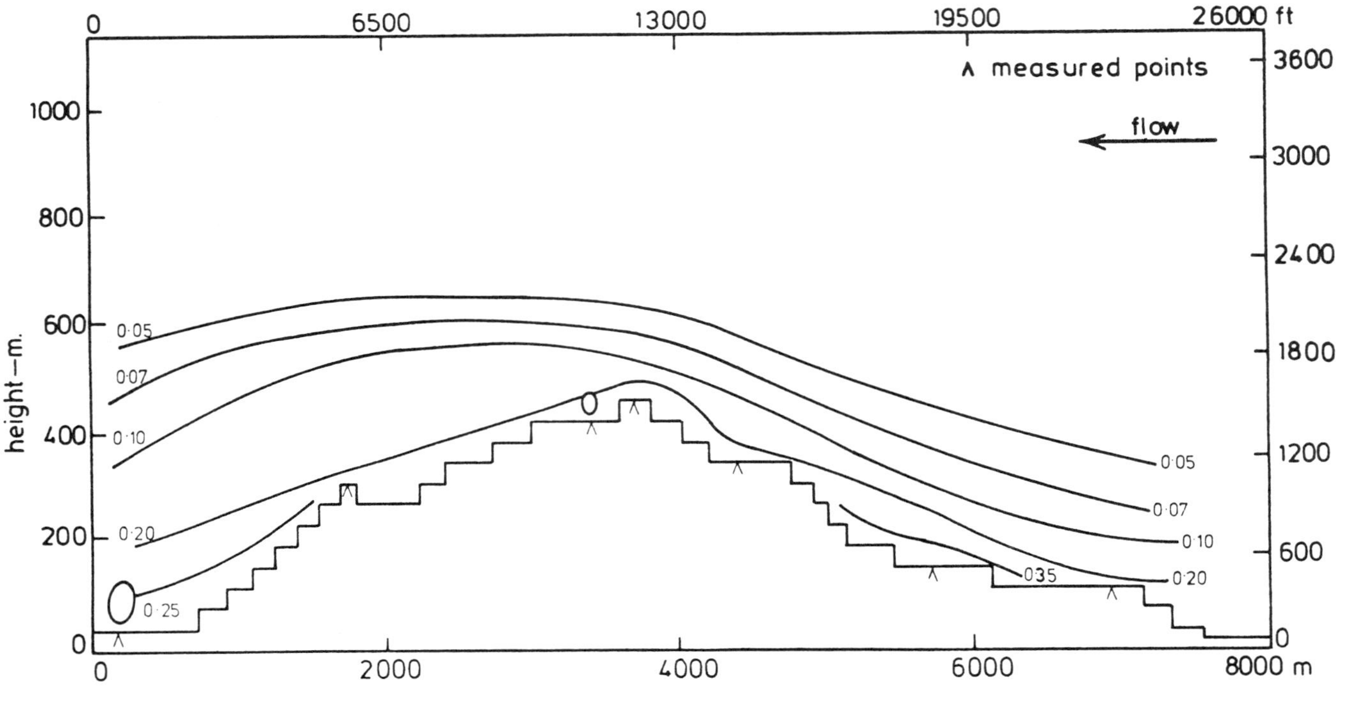

FIG 9b. ISOTURBS FOR MODEL B TERRACED, LONGITUDINAL CROSS SECTION YY

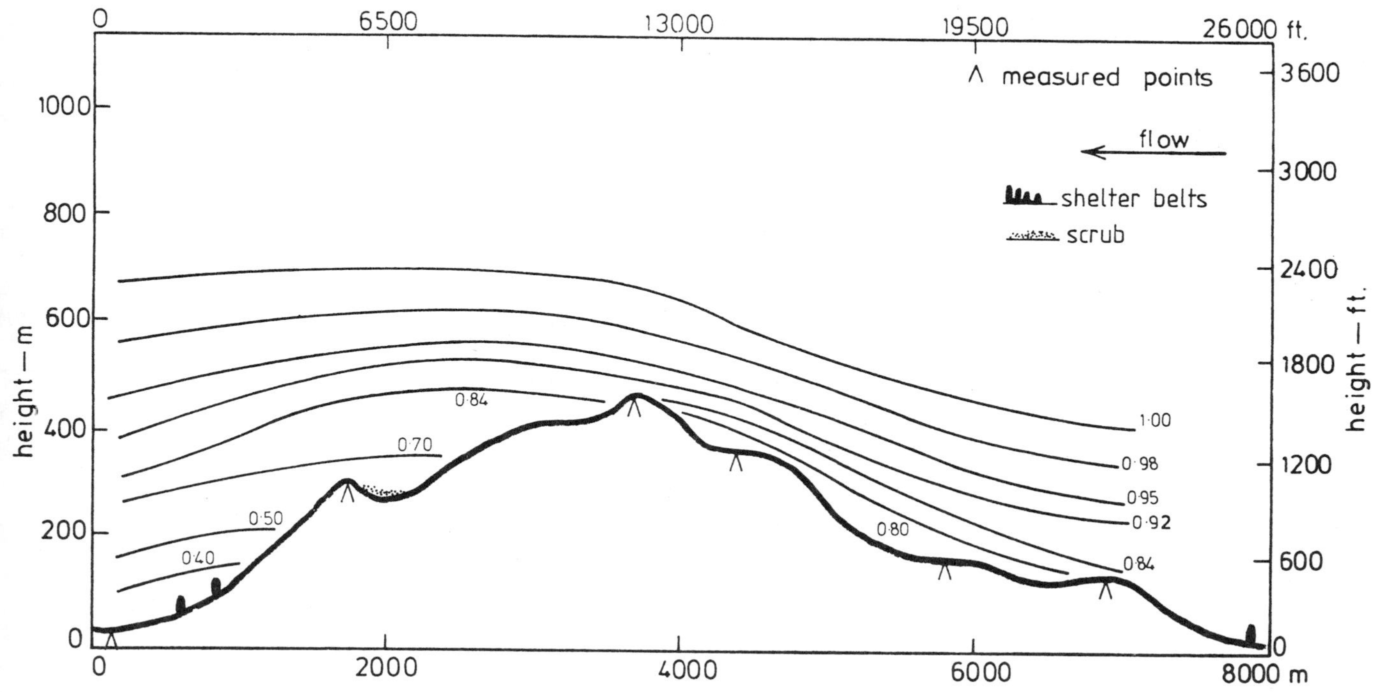

FIG 10a. ISOTACHS FOR MODEL B WITH ROUGHNESS, LONGITUDINAL CROSS-SECTION YY

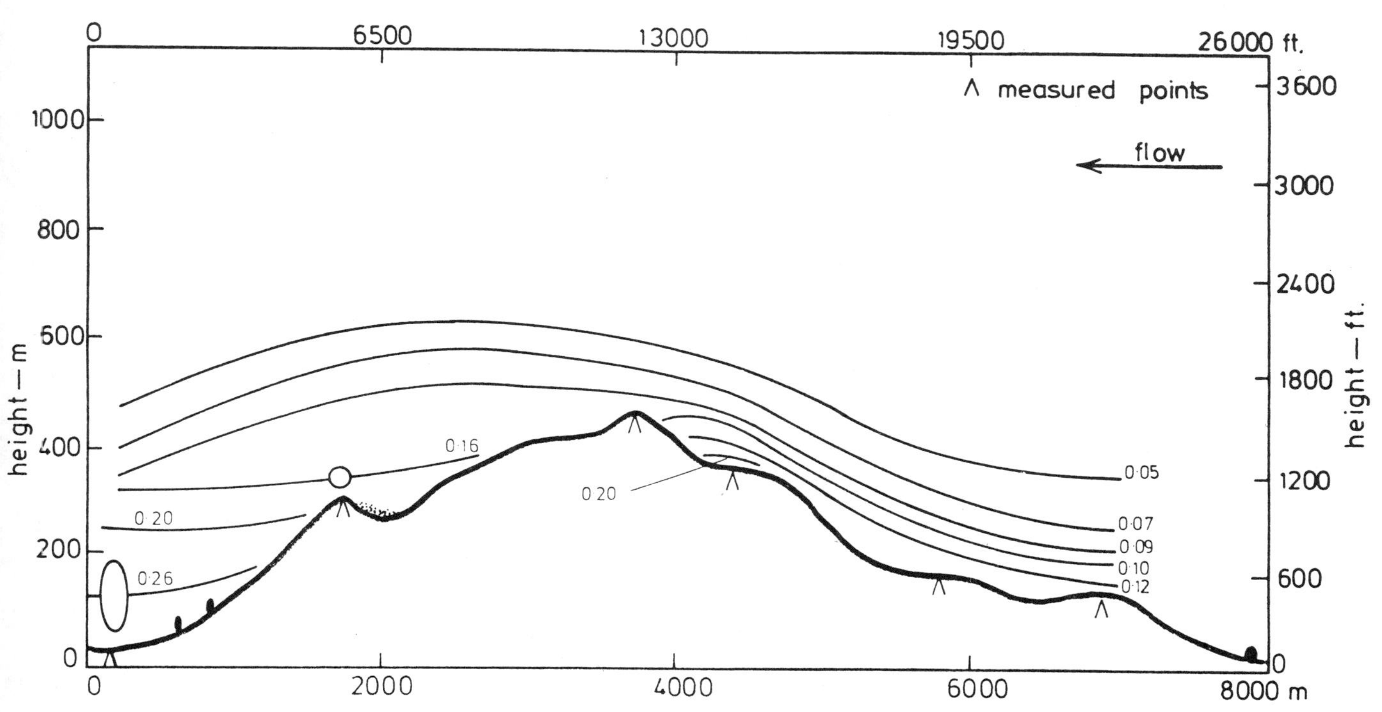

FIG 10b. ISOTURBS FOR MODEL B WITH ROUGHNESS, LONGITUDINAL CROSS-SECTION YY

3rd BRITISH WIND ENERGY ASSOCIATION WORKSHOP

LIST OF DELEGATES

ABBOTT, V. M.	Engineering Dept., Reading University, Reading, Berks. RG6 2AY.
AINSLIE, J. F.	CERL, Kelvin Avenue, Leatherhead, Surrey.
ALBERTO, V.	
ALCALDE, V. H. C.	Elec.Eng. Dept., Imperial College, Exhibition Road, London S.W.7.
ALEXANDER, A. J.	Mech.Eng. Dept., University of Technology, Loughborough, Leicestershire. LE11 3TU.
ALLAN, J.	Aircraft Designs (Bembridge) Ltd., Bembridge Fort, Sandown, Isle of Wight, PO36 8QS
ALLEN, J.	E.T.S.U., AERE, Harwell, Didcot, Oxfordshire OX11 ORA.
ARMSTRONG, J. R. C.	Taylor-Woodrow, 345, Ruislip Road, Southall, Middlesex. UB1 2QX.
ANDERSON, M. B.	Cavendish Lab., Madingley Road, Cambridge, CB3 OHE.
BAGULEY, M. J. G.	Holset Engineering Co. Ltd., 131, Parkinson Lar Halifax, HX1 3RD.
BANNISTER, W. S.	Mech.Eng. Dept., Napier College, Edinburgh, EH10 5DT.
BARNES, J. A.	Mech.Eng. Dept., Imperial College, Exhibition Road, London SW7 2BX.
BEDFORD, L. A. W.	ETSU, AERE, Harwell, Didcot, Oxfordshire, OX11 ORA.
BEURSKENS, H. J. M.	ECN, P.O. Box 1, 1755 ZG, PETTEN, The Netherlar
BOLTON, H. R.	Elec.Eng. Dept., Imperial College, Exhibition Road, London, SW.19 2BT.
BOYE, G.	Alternative Technology Group, Open University, Walton Hall, Milton Keynes, MK7 6AA.
BRAASCH, R.	Sandia National Laboratories, Division 4715, Albuquerque, New Mexico, 87185, U.S.A.
BRANNSTROM, A	Swedyards Corp, Box 416, S-401 26 Göteborg, Sweden.
BRAUN, K. A.	University of Stuttgart, ISD, Pfaffenwaldring D-7000 Stuttgart 80, W. Germany.
BROCKLEHURST, A.	AERO Dept., Westland Helicopters Ltd., Yeovil, Somerset.
BROWN, A.	James Howden and Co. Ltd., 195, Scotland Stree Glasgow.
BROWN, G.	Highlands and Islands Development Board, 27, Bank Street, Inverness, IV1 1QR.

BROWN DOUGLAS, G.A.	Mansfield, Chapelton by Strathaven, Lanarkshire.
BUEHRING, I.	Wind Power Unit, Chem.Eng. Dept., Exeter University, EX4 4QF.
BURNS, J. A.	Rotorway Composites Ltd., Tweed Road, Clevedon, Avon.
CLANCY, J. S.	Eng. Dept., Reading University, Reading, RG6 2AY
CLARE, R.	Sir Robert McAlpine & Sons Ltd., 40, Bernard St., London, WC1N 16G.
CLAYTON, B. R.	Mech.Eng. Dept., University College, Torrington Place, London WC1E 7JE.
COLLINS, R. J.	Vickers Ltd., Market Street, Eastleigh, S05 4FD.
COMYNS-CARR	Gifford and Partners, Carlton House, Ringwood Road, Woodlands, Southampton S04 2HT.
CONWAY, A.	Science and Industry Unit, BBC External Services, Bush House, Strand, WC2B 4PH.
COOPER, B. J.	GEC. Power Eng. Ltd., Whetstone, Leicester.
CURE, J. R.	
DALE	Northumbrian Energy Workshop Ltd., Tanners Yard, Gilesgate, Hexham, Northumberland.
DESTY, P.	Rampart Boat Building Co.Ltd., Vespasian Road, Bitterne Manor, Southampton, S02 4AY.
DESTY, R.	Rampart Boat Building Co.Ltd., Vespasian Road, Bitterne Manor, Southampton, S02 4AY.
DEVEREUX, A.	British Hovercraft Corp. Ltd., East Cowes, Isle of Wight. P032 6RH.
DIXON, J.	Technology (EM), Open University, Milton Keynes, MK7 6AA.
DOBSON, A. R.	Plessy Aerospace Ltd., Abbey Works, Titchfield, Hants. P014 4QA.
FORDHAM, E. J.	Cavendish Lab., Madingley Road, Cambridge.
FRAENKEL, P. L.	Intermediate Technology Development Group, 9, King Street, London WC2E 8HN.
FRANDSEN, S.	Physics Dept., Risoe National Lab., 4000 Roskilde, DK-DENMARK.
FREATHY, P.	Atkins Research and Development, Woodcote Grove, Asley Road, Epson, KT18 5BW.
FRERIS, L. L.	Elect. Eng. Dept., Imperial College, Exhibition Road, London. SW7.
FULLER, E. B.	A. B. Fuller Ltd., 196, Morland Road, Croydon, Surrey. CR3 9TL.

GAIR, S.	Elect.Eng. Dept., Napier College, Edinburgh.
GARRARD, A. D.	Taylor Woodrow Ltd., 345, Ruislip Road, Southall, Middlesex, UB1 2QX
GARSIDE, J.	Cranfield Inst. of Technology, Cranfield, Bedford. MK43 OAL.
GRIFFITHS, P. E.	Rutherford and Appleton Labs., Chilton, Didcot, Oxon.
HALES, R.	School of Mech. Eng., Cranfield Institute of Technology, Cranfield, Bedford, MK43 OAL.
HALLIDAY, J. A.	ERSU, Rutherford and Appleton Labs, Didcot, Oxon. OX11 OQX.
HANCOCK, M.	Wavepower Ltd., Carlton House, Ringwood Road, Woodlands, Southampton SO4 2HT.
HARRIS, R. I.	College of Aeronautics, Cranfield Institute of Technology, Cranfield, MK43 OAL.
HASSAN	
HERMANSSON, O.	Aerogenerator Dept., Kamewa, Sweden.
HOLMAN, G. S. F.	School of Engineering, Hull College of Higher Eduction, Hull.
HOLMES, B. A.	P. I. Specialist Engineers, The Dean, Alresford Hants.
HRYNISZAK, W.	The Energy Centre Ltd., Pennine House, 4, Osborne Terrace, Jesmond, Newcastle-upon-Tyn NE1 2NE.
HUGOSSON, S.	Malmstigen 15, S-18400 Åkebsberga, Sweden.
HUTCHINSON, A.	Electricity Supply Board, 27 Lower Fitzwilliam Street, Dublin 2.
IACOVONI, F	Comitato Nazionale Energio Nucleare, C.S.N. Casaccia, S.P. Anguillarese Km 1300, 00060 S.Maria di Galenia, Rome.
JAMIESON, P.	James Howden and Co.Ltd., 195, Scotland Street, Glasgow, G5 8PJ.
JANSEN, W. A. M.	DHV Consulting Engineers, PO Box 85, 3800 AB, Amersfoort, The Netherlands.
JEWELL, P.	17, Church Road, Southbourne, Bournemouth, BH6 4AS
JONES, J. M.	M.N.C. Products Ltd., Howgill, Newby, Penrith, Cumbria. CA10 3HQ.
JONES, K. C.	Hamilton Standard (UK Rep), C/o AEL, Station Approach, Bicester, Oxon OX6 7BZ.

KEAST, G. T.	Boving and Co.Ltd., Villiers House, 41-47 The Strand, London WC2.
KELLEDY, E.	Dept. Statistics, Trinity College, Dublin.
KETLEY, G. R.	British Aerospace Dynamics Group, Manor Rd., Hatfield, Herts. AL10 9LL
KILNER, J.	B.P. Research Centre, Chertsey Road, Sunbury-on-Thames, Middlesex. TW16 7LN.
KING, R.	Foster Wheeler Power Products, P.O. Box 160, Greater London House, Hampstead Road, London. NW1 7QN.
KINSELLA,	Dept. of Energy, Dublin, Ireland.
KITSON, R.	Davy-Loewy Ltd., Prince of Wales Road, Sheffield. S9 4EX.
KREIBIG, D.	Voest-Alpine AG, Dept. VSB14, Hessenplatz 8, A4020 Linz, Austria.
LEPOUTRE, F.	Cyclotex, 82, Rue d'Hem, 59170 Croix, France.
LIPMAN, N. H.	Rutherford and Appleton Labs, Chilton, Didcot, Oxon. OX11 OQX.
LOVETT, C.W.B.	15, Maple Close, Clacton-on-Sea, Essex.
LOWE, R. J.	Energy Research Group, Open University, Milton Keynes, MK3 6HN.
LUNDSAGER, P.	Test Plant for Small Windmills, Risø National Labs, DIC 4000 Roskilde, Denmark.
LYONS, R. A.	Atkins Research & Dev., Woodcote Grove, Ashley Road, Epsom, KT18 5BW.
McAULEY, J.	Dept. of Industry, Abell House, John Islip Street, London, SW1.
McHAMISH, G.	Boving & Co.Ltd. Villiers House, 41-47, The Strand, London, WC2.
McINTOSH, I.	School of Mech. Eng. Cranfield Institute of Technology, Cranfield, Bedford. MK43 OAL.
MAGUIRE, T. H.	Tom Maguire & Co.Ltd., Sunderland House, The Dockyard, Pembroke Dock, Dyfed.
MARSHALL, C.	Crown Agents, 4 Milbank, Westminster, London SW1P 3JD.
MASKELL, C. W.	ETSU, AERE, Harwell, Oxon.
MAYER, R. M.	New Technology Division, B.P. Research, Chertsey Road, Sunbury, Middlesex. TW16 7LN
MAYS, I. D.	Sir Robert McAlpine & Sons, 40, Bernard St., London, WC1N 1LG.

METHERALL, A.J.	Cavendish Lab., Madingley Road, Cambridge CB3 OHE.
MILBORROW, D.J.	CERL, Kelvin Avenue, Leatherhead, Surrey.
MOGHADDAM, T.	Mech. Eng. Dept., Imperial College, Exhibition Road, London, SW7.
MOORE, D.J.	CERL, Kelvin Avenue, Leatherhead, Surrey.
MORGAN, R.C.	Moorcot, Moortown, Nr. Tavistock, Devon.
MORRIS, C.J.E.	W.S. Alkins & Partners, Woodcote Grove, Ashley Road, Epsom KT18 5BW.
MOWFORTH, E.	Mech.Eng. Dept,. University of Surrey, Guildford, GU2 5XH.
MUNDY, C.	Performations (Magnets) Ltd., Chetney Manor, Swindon, Wilts, SN2 2PX.
MUSGROVE, P.J.	Engineering Dept., Reading University, Reading, RG6 2AY.
NANCE, C.T.	Medina Yacht Co. Ltd., Cowes, Isle of Wight. PO31 8BL.
NICHOLS, R	Aldborough Manor, Boroughbridge, N. Yorks.
O'DONNELL, I.H.D.	North of Scotland Hydro Electric Board, 16, Rothesat Terrace, Edinburgh.
O'FLAHERTY, T	Agricultural Institute, Kinsealy Research Centre, Malahide Road, Dublin 5.
OEI, D.J.	Energy Research Foundation, P.O. Box 1, Petten, The Netherlands.
OTWAY, F.O.J.	Generation Development & Construction Division, CEGB, Barnwood, Gloucester, GL4 7RS.
PAGE, D. I.	ETSU, AERE, Harwell, Oxon, OX11 ORA.
PAIRAUDEAU, R. D.	Rotorway Composites Ltd., Tweed Road, Clevedon, Avon.
PARKER, W. N.	Transmission & Technical Services Division, CEGB, Burymead House, Portsmouth Road, Guildford, GU2 5BN.
PARNELL, A.	18, Fetters Row, Edinburgh, EH3 6RH.
PEDERSEN, B. M.	Fluid Mechs. Dept., Technical University, Building 404, 2800 Lyngby, Denmark.
PETERSEN, G.	Institute fur Physik, Postfach 1160, 2054 Geesthacht, W. Germany.
PHILLIPS, D. C.	Materials Development Division, AERE, Harwell, Oxon. OX11 ORA.
PIRRIE, N. D.	Sir Alfred McAlpine & Sons Ltd., 40, Bernard St., London WC1N 1LG.

PLATTS, M. J.	Gifford & Partners, Carlton House, Ringwood Road, Woodlands, Southampton SO4 2HT.
PONTIN, G. W-W.	Wesco Windmills Ltd., Iroho House, Bolney Ave; Peacehaven, Sussex. BN9 8HF.
POWLES, S. J. R.	Cavendish Laboratory, Madingley Road, Cambridge.
PRETLOVE, A. J.	Engineering Dept., Reading University, Reading, RG6 2AY.
PRIEST, H.	Davy Lowy Ltd., Prince of Wales Road, Sheffield, S9 4EX.
PYBUS, D.	The Torrington Co.Ltd., Yarm Road, Darlington, Co.Durham. DL1 4PP.
REVELL, P. S.	Dept. of Engineering Science, University of Warwick, Coventry, CV4 7AL.
RHODES, H.	NEI Clarke Chapman Cranes Ltd., Smith House, Town Street, Rodley, Leeds, LS13 2TG.
RICHARDSON, J.	Balfour Beatty Power Construction Ltd., 7, Mayday Road, Thornton Heath, Surrey. CR4 7XA.
RIDDELL, J. C.	J. C. Riddell & Associates, Aroyl House, Hollybush Green, Collingham, Wetherby, Yorks. LS22 5BE.
ROCKINGHAM, A.	Planning Dept., CEGB, 20, Newgate Street, London EC1A 7AY.
RODWELL	Trinity Lighthouse Service, Tower Hill, London, EC3 N4 DH.
ROWBOTTOM, R. F.	Midland Bank Ltd., 27/32, Poultry, London EC2.
SAYERS, K. H.	Capricornia Institute of Advanced Education, Rockhampton, Queensland, Australia.
SCOTT, J. Y.	James Howden & Co.Ltd., 195, Scotland Street, Glasgow.
SEXON, B. A.	Engineering Dept., Reading University, Reading, RG6 2AY.
SHARPE, D. J.	AERO Eng.Dept., Queen Mary College, Mile End Road, London E1 4NS.
SINCLAIR, I.	I & M Sinclair Electrical Ltd., Bridgend, Thurso, Caithness.
SOMERVILLE, W. M.	NEI Clarke Chapman Engineering Ltd., Clarke Chapman Marine, Victoria Works, Gateshead, NE8 3HS.
SOPER, B. M.	ETSU, AERE, Harwell, Didcot, Oxon OX11 ORA.
STACEY, G.	Engineering Dept., Reading University, Reading RG6 2AY.
STAMBOLIS, C.	Heliotech. Associates.

STEVENSON, W. G.	North of Scotland Hydroelectric Board, 16, Rothsay Ter, Edinburgh.
SWIFT, R.	Technology (EM), Open University, Milton Keynes, MK7 6AA.
SWIFT-HOOK, D. T.	CERL, Kelvin Avenue, Leatherhead, Surrey.
SYRETT, J. J.	CEGB Research Division, Courtenay House, 18, Warwick Lane, London, EC4P 4EB.
TAYLOR, D.	Alternative Technology Group, Open University, Milton Keynes MK7 6AA.
TAYLOR, G. J.	CERL, Kelvin Avenue, Leatherhead, Surrey.
TAYLOR, R. H.	CEGB, Planning Dept., Laud House, 20, Newgate Street, London, EC1A 7AY.
TAYLOR, R. J.	ETSU, AERE, Harwell, Didcot, Oxon. OX11 ORA.
THOMSON, R. B.	Clayton Son & Co.Ltd., 23, Hodgson Avenue, Leeds, LSI7 8PH.
TODD, R. W.	Centre for Alternative Technology, Machynlleth, Powys, Wales.
TULLEY, A. A.	Science Museum, Exhibition Road, S.Kensington, London SW7 2DD.
UCELLI, G.	Calzoni, Bologna, Italy.
VAAHEDI	
VALTER, G. P.	E.C.N., PO Box 1, 1755 ZG Petten, The Netherlands.
WALKER, J. F.	CEGB, Marchwood Engineering Labs., Marchwood, Southampton SO4 4ZB.
WARNE, D. F.	E.R.A. Technology Ltd., Cleeve Road, Leatherhead, Surrey.
WATERTON, J. C.	South of Scotland Electricity Board, Cathcart House, Spean St., Glasgow G44 4BE.
WATSON, G.	Northumbrian Energy Workshop Ltd., Tanners Yard, Gilesgate, Hexham, Nthbld.
WHEATLEY, J. H. W.	National Maritime Institute, Feltham, Middx.
WHITTLE, G. E.	Engineering Dept., Reading University, Reading, RG6 2AY.
WIELAND, K.	VFW Bremen, Hunefeld Str. 1-5, 2800 Bremen, W. Germany.

WILLIAMS, P.B.	John Laing Design Associates Ltd., Page Street, Mill Hill, London, NW7 2ER.
WILSON, D.	Cavendish Lab., Madingley Road, Cambridge.
WILSON, R. R.	James Howden and Co.Ltd.,195 Scotland Street, Glasgow.
WINTER, A. J. B.	Computer Planning Services, 130c Broadway, Didcot, Oxon OX11 8UB.
WORTHINGTON, P. J.	CERL, Kelvin Avenue, Leatherhead, Surrey.
YEO, K. L.	Rutherford & Appleton Labs., Chilton, Didcot, Oxon.